SCRAP

Also by Ernest Dudley

FOR LOVE OF A WILD THING
RUFUS: THE STORY OF A FOX
ARTHUR THE TV CAT
RANGI, HIGHLAND RESCUE DOG
CHANCE AND THE FIRE HORSES

SCRAP
The Gentle Wildcat

Ernest Dudley

FREDERICK MULLER

First published in Great Britain 1974 by
Frederick Muller Ltd., Edgware Road, London NW2 6LE

Printed in Great Britain by
Clarke, Doble & Brendon Ltd, Plymouth

ISBN 0 584 10249 6

With the Author's
Grateful Thanks to
I. Neil Macpherson and his fellow-Directors;
Eddie and Joanne Orbell, and Colin Burton
of the Highland Wildlife Park,
for their very kind help
over the telling of Scrap's
Story

Photographs by
Colin Burton, Ian Bradshaw, Geoffrey Kinns, Jim Taylor and
The Scottish Daily Express

Contents

Foreword

The wildcat was here long before us. A pre-mammoth animal, the wildly beautiful Scottish Highlands were its rightful home, its natural habitat. Because it was a brave, savage creature, men hunted it for sport until it became almost extinct; and it was enabled to find a last refuge only in the lonely Highlands. It is still hunted here as vermin, accused of preying on lambs and game, though there is some doubt as to whether the wildcat is as guilty as claimed.

In the mists of pre-history the wildcat was adopted by Highland clans, the predecessors of the great Clan Chattan, themselves renowned for their warlike, savage prowess in battle. There is no written record of when the Highlanders adopted the wildcat, but during AD 200 the Highlands were the military centre of the Pictish warriors who made up the two leading clans, the Orcs (the Boars) and the Cats (the Cat Clan), so that it may be presumed that Scrap's ancestors had long before that time inspired the fighting Highlanders; and the wildcat had become one of the emblems they wore when they ranged the glens and hills of Badenoch, leaping to the attack against their foes.

These were the ancestors of the legendary Gillichattan Mór and Muriach, the Prior of Kingussie (from whom the Macphersons are descended), and their sons and successors, who continued to feature the wildcat as their chief's crest. Fierce in defence of its lair, its home; independent in spirit; prepared to fight to the last for all that it holds dear – what could provide the Macphersons with a more appropriate symbol?

The wildcat became part of Scotland's written history in 1672 when it was shown in the Lyon register as Cluny Macpherson's personal crest, full face and defiant, with claws "ungloved"; with its famous motto, "Touch Not the Cat But a Glove", meaning "Touch not a cat unless you've got gloves on", warning aggressors that those ungloved claws can strike.

To Macphersons the world over the wildcat remains an example of the freedom and wild beauty of the Highlands, epitomising the Highlands' warrior courage; and it is these great qualities that Scrap's fellow wildcats still exemplify today.

William Macpherson of Cluny
27th Chief of the Clan Macpherson

On even the distant approach of a stranger two little Jekylls became two little Hydes with flattened ears and glaring eye-slits, and if the visitor left without blood being drawn it was owing to his belief in discretion being the better part of valour . . . I doubt if there is any other mammal of its size and weight capable of inflicting such wicked damage on an aggressor.

GEOFFREY FISHER, EDINBURGH ZOOLOGICAL GARDENS

Several well-known naturalists are endeavouring to keep wild-cats, but even the litters brought up by hand are still proving vicious . . . Many attempts have been made to tame a specimen, but either the kittens succumb to illness . . . or, if they have lived, have failed to respond to kindness . . . This species is noted for its strength, agility and fierceness.

GRACE POND, FZS

This animal is the fiercest and most destructive animal we have. Its strength and ferocity when hard pressed are perfectly astonishing.

DR MAURICE BURTON, FZS

It is easily the fiercest of our larger animals. No amount of handling or soft words counteracted the inborn ferocity of their natures.

ROWENA FARRE

The cat became transformed; almost, I had said, transfigured. The last trace of resemblance to a fluffy Persian kitten vanished utterly; in its place was a noble, savage wild animal at bay before its ancestral enemy. Laying his ears not back but downward from the broad flat skull, so that the very tips and the tufts of the hair that grew from within them were all that turned upward, baring every fang and gum in his head so that the yellow eyes became slits of rage and hate, swelling his ringed tail to twice its previous girth, he reared himself back . . . One paw was lifted with extended talons . . . the forelegs seemed to have elongated like telescopes; those velvet limbs had in an instant changed from instruments of locomotion into long-reaching weapons to rake and slash. As an image of primordial ferocity I had seen nothing to equal it; it was splendid, it was magnificent, but it was war.

GAVIN MAXWELL

PART ONE

Chapter One

It had rained the past couple of nights, bright and sunny during the day, and then rain again as the evening fell. And even though it was mid-May this rain was chill, slashing into your face so that you slitted your eyes against it. It was past midnight, the sort of night you would say was poacher's weather.

The wildcat hated it. She was a female wildcat, and she was heavy with young; they were stirring violently in her belly and soon it would be her time. This was another reason why she remained in her lair, coming out only to hunt for food, returning as quickly as possible, shaking the gauze of rain-mist off her whiskers to pick her way carefully through the soaked grass and mud as she headed back home.

Suddenly, she stopped. Every line of her tensed, one paw raised in mid-air, her striped tail flicking very slightly from side to side. Her tail almost didn't move, now, only two inches of its tip barely quivering. Her nose flickered as it picked up the smell of steel, and that other only too familiar smell. Man. Only a few feet away from her was, she knew, a gin-trap, its dreadful jaws hidden by scattered earth. Satisfied that she knew exactly where it lay, that the trap was not, after all, in her path, she relaxed. She put the raised paw down, slowly, gingerly, and then she moved on.

She was making her way back to the edge of Badenoch Forest. The moon appeared and disappeared among the black clouds still swollen with rain. For a moment her eyes glowed greenish gold in the shadow of a clump of bushes. The star of white fur at her throat and the second larger star below broke up her shape. Her form seemed to disintegrate before your very eyes. A sharp breeze blew from the direction of Loch Insh, so that the branches of the larches above her scraped noisily against each other and raindrops spattered her.

She had made kills of mice and voles, and had been stalking an unsuspecting rabbit sheltering among some bracken when the movements in her belly had become more urgent, and so the rabbit had lived. On the edge of the forest she paused, raising her head to scent the wind. Satisfied that no danger threatened she slipped through the undergrowth along the forest edge.

Now her lair was only thirty yards away. She paused, shook the rain from her shoulders in a fine spray, and again sniffed the wind off the loch. It was carrying with it the scent of otter. Her nose wrinkled; she drew back her lips so that her white upper canines glistened in a scowl. She swung to her right, back into the forest, to approach her lair indirectly. Cautiously, treading as lightly as she could, the weight of her unborn kits slowing her, she moved like a shadow, her black tiger-like stripes rippling against their tawny background.

She gained the familiar clump of pine-trees near her lair. She was only a few feet from safety. She paused and crouched, then with her powerful hind legs she jumped a distance of several feet, landing only a few inches from the lair's entrance. This way she had left no trace of her paws, the scent of which could have been picked up by hunting dogs. She made no pause at the entrance but went straight in between the roots of a huge oak-tree which had been so badly storm-battered that, although it had survived the driving winds and blizzards, it had been pushed back and its roots exposed. Within these roots the wildcat had found a cavern large enough to accommodate her family.

The wildcat and her mate had found the place several weeks

before. They had occupied the lair together until the last week when the male wildcat, aware that his mate would be giving birth to the kits, had made off. He had no interest in the birth of the kits he had fathered; in fact, he would almost certainly have killed them at birth.

The female wildcat was on her own. To hunt for herself, to keep watch herself against the threat of her enemies. She was young, only a year old, and this was to be the first time she would bear kits. She knew instinctively what lay before her. She was not daunted by the absence of her mate, she had no fears for the future or for the future of her offspring.

The lair went some six feet into the earth and she and her mate had already scraped out enough earth to give a space widening the cavern until it was about four feet square. The wildcat stood in the centre of the cavern for a few moments, eyes fixed on the entrance, nose lifted to catch any scent that might reach her from the night outside, ears cocked against any sound that might trouble her. There was nothing. For a few moments the darkness of her lair was lightened as the moon appeared from behind a bank of cloud and some of the moonlight seeped into the cave.

She heard the sudden, distant miaouw of her mate, a harsh screech that turned her face into an instant mask of fury, and a hissing snarl escaped her drawn-back lips, as if warning him from the darkness to keep his distance; she knew the deadly danger he meant to her kits. That miaouw once again, further away, trailing into a thin, high note. This time her body stiffened, her snarling scowl softened. Was her mate signalling her a warning of danger? She moved towards the lair's entrance, her nose lifted, her ears swivelled forwards to catch the slightest sound. For several moments she paused there, still and taut. There seemed to be nothing alarming in the wind. There was no further miaouw from her mate. Then the moon sped behind dark banks of cloud once more, and once more the lair was pitch black.

The wildcat relaxed; the darkness was her friend. Here a

sense of security filled her. Muscles, sinews and nerves unwound. She moved round slowly in her own length, her large paws smoothing the earth. There was no sign of any nest being prepared by her, only a few strands of grass and some pluckings of fur which she had torn from around her teats. Slowly she lay down, spreading herself easily on the floor of the cavern, curving her body. She began to tense: her short thick tail lay straight, flickering at the tip. Her eyes were calm, half-closed. She was breathing a little heavily but that seemed to be the only change. What followed seemed automatic, a matter entirely of reflex action. What were her thoughts, as her kits were born one by one? Inside that strong-looking, slightly flattened skull, what memories of the past were handed down to her genetically over thousands upon thousands of years?

Each kit was born head first. There were only a few minutes in between the first and second and third kits. As each tiny shape lay beside her she chewed the umbilical cord, her rough tongue licking the wet, supine body. The fourth and last kit was born nearly fifteen minutes after the third. Then came the afterbirth: she licked up the last trace of it, tension rising in her again, for she knew she must remove any sign or scent of it. Instinctively she knew that it might otherwise have betrayed her to an enemy. She pawed the four tiny shapes into the warmth of her belly, her hind legs curving round them as at once they began searching for her exposed teats, and she began suckling them.

She stretched out, eyes half-closed again, purring.

Her four kits, a female and three males, were quite blind. In colour they were lighter than their mother, a pale gingerish tawny colour, the tiger stripes a light brown. Still purring as she continued to suckle them, she began licking each kit carefully, removing any possible trace of the afterbirth, and her rough tongue seemed to energise each tiny creature with a kind of life so that even as they sucked at her milk they mewed and squeaked.

And from the very moment of her birth, the female kit,

Scrap, the name by which she was to become known, began learning to use her sense of hearing, her sense of smell and of touch. From the very beginning she was the recipient of signals from the tiny world about her; she could feel her three brothers push against her as they, too, sought her mother's warm belly and the sweet-tasting milk from her teats.

Hers was a world of half-dreaming, half-waking, her innate curiosity only vaguely aroused for the first hour after her birth while she lay supine and her mother cleaned her with that ever-ready rough tongue that warmed and comforted her and at the same time massaged her tiny muscles, setting her blood and nerves tingling. She squeaked and mewed, as she heard her brothers squeak and mew. She lay very quiet as her mother's deep, throaty purring reached her, filling her tiny being with the knowledge that she was safe and secure.

Her mother's paw caught her to her belly, and once more Scrap's tiny mouth nuzzled into the furry warmth, searching for that sweet-tasting milk.

For a few moments the lair lightened again, as the moon reappeared from behind a black mass of cloud. The wildcat paused, her body tensed, though Scrap and her brothers were still too busy suckling her to notice her tension. Now her eyes were wide open, lids drawn back, her ears swivelled forward as she watched the entrance to the lair, tensed to catch any sound outside. She was no longer purring.

The moonlight faded, all was dark again and once more the wildcat's tension relaxed, she began purring as she licked each kit all over again.

And Scrap squeaked and mewed, the milk dribbling from her mouth, and then she fell asleep.

Chapter Two

The wildcat's mate, father of the four kits she had earlier given birth to, proceeded along the forest path. On a capful of the dawn wind he caught a scent which jerked his nose into the air. The smell of steel. Steel and the scent of human beings. It was the same scent he had sniffed in exactly the same place the past several weeks. The smell of steel his mate had scented. His eyes were drawn back into narrow slits by his striped mask of a face. His jaws were opening in a scowl, his canine teeth glistening. His whole body stiffened. His tail flicked as he crouched there, staring intently ahead of him. His flat-topped skull was moving slightly from side to side as the scent reached him and he collected the information which told him that only two or three feet away was the trap. The steel gin-trap. You might think there is something romantic about the idea; an exciting picture of moonlight poachers setting their snares and traps in the shadows grips your imagination. It isn't quite like that.

As the wildcat advanced more slowly he could smell not only the steel and human hands, he could smell a hare, and its blood. It was not the first time he had encountered a gin with a victim struggling in agony to extricate itself from the cruel jaws. He heard the rattle of the chain which held the gin to a pin driven into the ground. It was a sound he had heard before. The hare was kicking its hind-legs in the air, its long teeth clicking as it wriggled in agony, rolling its eyes.

The wildcat saw that its front-paws gripped in the trap were already oozing blood and swelling as the hare tried to extricate itself. Should he take the opportunity to kill it for himself? He hesitated. It was a tempting sight, this helpless victim apparently all ready to be killed without having the trouble of hunting it. It was a big hare and it lay in a spatter of blood and saliva from its lacerated tongue and gums where it had tried to bite the chain apart. Now it sprawled, utterly exhausted; now was the moment to move in quickly for the kill. Still the wildcat continued to sniff the air, eyes narrowed with suspicion.

The hare gave a gigantic twitch, almost rolling over on its back as it made another effort to free itself. There was more blood and at the smell of the fresh blood the wildcat's own jaws dribbled with saliva at the prospect of this tempting meal. He put forward a paw very tentatively. Then he pulled back. He was still too scared to get too near that smell of steel which spelled death to the hare and could mean equal danger to him.

A blackbird which had been watching with intent gaze the scene below from a nearby branch suddenly let out a series of harsh, nervous notes. The wildcat recoiled; it was as if the blackbird had sounded a warning that he'd better leave the hare alone. The wildcat scowled and hissed at the blackbird, which took off, its wings flapping erratically as it made its way through the trees and disappeared. The wildcat gave the hare another glance and then, as if shrugging its shoulders, moved on. He ignored the hare's more frenzied threshing, and left it to its agony.

He found himself on the edge of a clearing where a flock of jackdaws were feeding. The wind was blowing in his face and his nose wrinkled as the pungent smell of the jackdaws reached him. He crouched, trying to make up his mind whether to make an attack or not, then one of the jackdaws flew into the branches of a nearby larch-tree. The wildcat watched it with concentrated interest. The jackdaw started to preen itself and the wildcat knew then that it had not given an alarm-signal to the others. His presence had not yet been noticed. He looked at the other jackdaws and they were feeding casually. He crouched lower on his belly and began to edge himself towards them. All the time he kept one eye on the jackdaw in the larch which was still preening itself, drawing its flight feathers through its bill with a single quick pull. Suddenly it stopped and the wildcat knew at once that it had seen him. The jackdaw squawked, *kow-kow-kow*, and flew off into the forest. At the same time the flock took off and followed it, setting up a great clamour with their screeching *kow-kow-kow*. The wildcat made a savage rush at the last remaining jackdaw on the ground, lunging at it with a paw, claws glinting. But it was too late; with a clatter of wings and a *kow-kow-kow*, the jackdaw sped off after its fellows.

The wildcat moved through the undergrowth for twenty yards until he emerged again on to the moss-grown path, along which he made his way as before. He kept to the shadow of the bracken that spread out along the path's edge. Inevitably, even as he scented them, the birds and other animals in his vicinity were scenting him. The birds cried their alarm and flew helter-skelter through the forest, the rabbits and other small creatures, voles and mice, all knew that he was on his way and all put as great a distance as they could between them and him.

He paid no attention to the commotion he was causing all around him during his progress out of the forest towards Loch Insh. You would expect to be noticed with such alarm and terror-stricken awe. This was part of the scheme of things This was the rule of the game. Every animal of the forest, every bird, looked after its own interests and those of its mate and its young. And in order to protect themselves, in order to survive, which was their object in being there at all, they had to make sure that while hunting for food they themselves were not overtaken by another predator and became in turn a victim.

The wildcat came out of some ash-trees, and there across the pathway some ten yards ahead was something which caused him to stop once more. He stood rigid. The scent assailed his nostrils and, as he tipped his face to one side, the smell of food caused the saliva to drip from the side of his jaw. It was the smell of a farmyard hen. He crouched, his body tense, every muscle coiled to impel him in a spring. And then, just as he was about to jump, he caught another scent, very different from that of the fowl that lay there. This time it was the strong musty scent which he identified as a different bird. A grey-backed, hooded crow.

The two scents merged together and it was this that gave the wildcat pause to consider what the information he was receiving meant. The tension of his muscles relaxed and, slinking very low on his belly, he edged himself forward inch by inch. A rabbit – surprised by his unexpected approach – lifted itself onto its hind-legs and bounded off, scut dazzling white in the dawn light, eyes bulging with fear. As it disappeared into the forest, the wildcat moved on, his nose sniffing the air all the time as the two smells of hen and hoodie grew stronger. And now there was another, fainter, smell. One which was entirely alien to him. Entirely new. A scent he had never sniffed before.

Now, as he was only a few feet from the two shapes he stood up on his hind-legs for a few seconds and saw clearly what lay before him. The hen spread out, its feathers ruffled, a large white fowl, its eyelids closed in death. And beside it lay the greyish shape, the black wing-feathers glinting. It was the hoodie giving its last kicks of life. It had plunged its beak into the hen's body and had swallowed the cyanide planted there by human beings who had used the hen as a trap to catch the wildcat. The wildcat sniffed, and because of the smell of man instinctively knew what the trap meant for him. He hissed and scowled and moved on.

He came out of the forest, paused on the edge for a few moments nosing the wind. He darted across the road and ran along it until he came to the bridge of the railway-line to Inverness. He turned down under it. He paused in the pinkish, pearly light of the mist rising off Loch Insh, and sniffed the tang of fresh water, then proceeded slowly down towards the lochside shore.

He traversed the width of grey sand and scattered rocks until he reached the water's edge. He sniffed the surface of the water and then tapped it with his right fore-paw. He watched the drips of water fall from his paw back into the water, causing little ripples. He sniffed the surface of the water again, lapped a few mouthfuls, and then made his way along the edge of the water, jumping aside as the early morning breeze swept a few ripples against his paws.

He turned back as if to survey his paw-marks along the grey sand and then went on until he reached a spur of shore which jutted out into the loch. The loch was fed by the River Spey which brought with it the brown trout. A red-breasted merganser drake with its dark-green double crest all untidy in its flight took off from a clump of reeds on the edge of the sandy spur, crying its alarm. The wildcat watched it as, its red breast glinting in the hint of the early morning sun just beginning to peer over the horizon, it swirled and tilted, then disappeared towards the other side of the loch.

Further along, where the spur rejoined the main shoreline, there was a rustle in the water-weed and a grebe rose from its nest with a squawk and flew off in the direction of the red-breasted merganser. The wildcat turned his attention to the grebe and watched it as it, too, circled overhead and disappeared above the mist and lost itself towards the other side of the loch. Although the waves against the shore were small, they were nevertheless ever-lapping where shore-weed and quillwort and the water milfoil grew.

Among these plants lived the plankton and other organisms and molluscs which supplied the trout with their food. And the water was also enriched for the trout by the birch and alder leaves falling into the burns that found their way into the loch, and it was the brown trout which interested the wildcat.

By now he had reached the very tip of the spur and stared out across the surface of the water. A few yards ahead, where the mist lay in swirls of cottonwool white, he caught a flicker of a ripple and at once his body tensed and he crouched, his ringed tail flicking very slowly to and fro, his eyelids drawn back so that his eyes were glittering greenish slits. There was another ripple in the water ahead and in that instant the wildcat launched itself, rushing forward for two or three feet and then taking off with his powerful hind legs.

His forepaws hit the water before the rest of his body and his claws were already caught in the wriggling body of an eighteen-inch-long brown trout. His shining teeth bit through the glistening scales and, as he turned with the fish in his mouth, he used his front paws to help propel himself back to the spur of grey sand. He threw the fish down and began eating it as fast as he could, his teeth crunching into the bones and flesh of the trout.

The wildcat was hungry and little remained of the fish as he turned and made his way back up to the shore and then along to where the path ran underneath the railway bridge.

He was still licking his jaws with his long pink tongue as he came near the oak-tree beneath which he knew his mate lay with her four kits. He paused, head up and tail straight behind him moving slightly as he listened and sniffed the air. Very faintly he could hear the mewing and squeaking of the kits down there in the dark lair. He heard his mate's hiss of warning anger, he caught the scent of her and the new scent to him of their offspring.

He turned aside and made his way deeper into the forest.

Chapter Three

Scrap opened her eyes after ten days of lying in the dark. Like her own, her brothers' eyes were open, and like hers were bright blue in colour. She was never to know this for, again like them, Scrap was colour blind. The blueness of her eyes really meant absence of pigment. As the days went by the blue "non"-colour, as you might describe it, would give way to the characteristic, beautiful greenish-gold, smoky-amber colour of a wildcat's eyes.

Slowly Scrap began to make out the three shapes of her brothers as they lay close together in the darkness. Like them she was beginning to focus, and she peered into their faces as they peered back at her. Already all four pairs of eyes were filled with puzzled inquisitiveness.

But if she was only just learning to make use of this new sense of sight, her other powers, of hearing and of scent, were in full working order. They had been from the moment of her birth. Scrap joined in with her brothers' squeaking and mewing as she caught the sound of her mother's anxious purring at the lair's entrance, and at the same time scented her.

She was gorged with a rabbit, plus mice and voles which she had caught, and in her mouth she carried a large vole which she proceeded to eat while she lay down, Scrap and the other three immediately nosing into her belly, searching for her teats.

One of her brothers had already established himself as the

strong one of the quartette, always pushing the others, Scrap included, aside, when their mother suckled them. But Scrap insisted on coming after him, and in her turn pushed her two brothers aside as she reached her mother's teats. She had accepted her place in the pecking order of the family, second to that of her stronger brother, while below her in the order were the other brother and the fourth kit, the weakling who was always pushed aside at feeding time, trampled on quite ruthlessly, his whimperings and squeaks of pain ignored.

Scrap never whimpered or squeaked when her strong brother pushed her aside – on the contrary, she always fought back. And though she always lost to him she nevertheless never failed to put up a fight against his bullying and bit at him as best she could with her little milk teeth.

The world of darkness in which she had lived since her birth up to now, before her eyes were open, had been a world of scents and sounds. The warm scent of her mother's fur, the warmth of her three brothers as they struggled beside her to feed; their squeaks and mewing. And the deep harsh purring of her mother lying there giving them food.

Sometimes a sound would reach Scrap from the world outside, the world which she was yet to discover. Her mother's purring would stop and she would give a low warning guttural growl. Her body would stiffen with apprehension and Scrap and her brothers followed her example. With them, she listened intently to the sound, although she could not make out what it was. She waited for her mother's growling to subside, for her throaty purring to start again as she began once more to suckle them.

Apart from her mother's and her brothers' scent and vague scents which drifted in from outside the lair and the sounds which alerted her mother, Scrap experienced only one other sense, that of touch.

She made direct contact with her mother when she kneaded her tiny paws into the teats' mammalian area to help the milk secretion to flow. She would nuzzle her mother's belly with her forehead, that part where there was the dip or depression between the top of her nose and her eyes. That sensory pit, a special sense organ, informed Scrap that she was hitting the spot. Warmth, her mother's milk, security. Love, if you like.

Together with her brothers she would squirm and wriggle and mew as her mother cleaned her with her rough tongue, licking her entire tiny body several times a day to make sure that she was perfectly clean. Scrap would feel her brothers push against her as they too squeaked and mewed as they were given the same treatment. And against her fur and skin there was the earthy feeling, warm and smooth, of the floor on which they lay.

From the moment of her birth Scrap had learned to use her sense of hearing and her sense of smell to provide her with information of what was going on in the tiny world about her. Soon there came hints of that other world outside. Hers was a world of half-dreaming, half-waking, sometimes fully alert, sometimes half asleep, but all the time information was reaching her and instinctively she was storing it up in her tiny brain. Her mother's purring sounds, then her harsh warning notes, all these Scrap soon began to interpret accordingly.

There was danger when her mother's body tensed and when the purring turned to a rasping growl. Relaxation and the sense of well-being and enjoyment filled her tiny body when her mother relaxed and purred and licked her with her warm, rough tongue. And Scrap sensed that her three brothers experienced the same sensations, though the fourth weaker one seemed a little slow in the uptake. She and her two other brothers took advantage of this, pushing him aside when it came to the matter of reaching her mother's teats.

Every day there were new smells for Scrap to scent, new sounds to hear and to tuck away in her memory against the time when she would hear them again and recognise them for what they were. The moan of the wind off the loch outside the lair. And that wind bringing with it the scents of the waters of the loch, of the trees and the flowers. And that same wind brought the rustle of the leaves of the trees outside, the squeaking of the branches as they swayed in the wind that blew through the forest.

One reason why Scrap, together with her brothers, had been born blind was that without the power of sight you are compelled to rely on the senses of hearing, scent and touch. For these senses will be of most importance to you as you grow up; your sense of sight would never be as efficient as those other senses. This was one reason why Scrap's eyes remained closed for ten days after she had been born. The other was simply that if she couldn't see she wouldn't attempt to stray from her mother and the lair. What you couldn't see, you couldn't be curious about. Not being able to see meant she wasn't attracted to anything outside her immediate vicinity, the warmth and scent of her mother and the warmth and scent of her brothers.

Her curiosity was only vaguely aroused for the first few hours after her birth. Her mother had cleansed her with that ever-ready rough tongue; and, as her innate curiosity increased, it was satisfied for the time being by the different scents and sounds that reached her and the information that touch was giving her, the touch of her mother's paws, the touch of her brothers' tiny bodies, the sense of security that it all contributed to her dark world. All these tiny bits of information, which reached her like pieces of a jigsaw puzzle, she fitted into her memory-store. The warmth and sensation of being protected, broken every now and again by the sense of alarm that flooded every fibre of her mother's being and those of her three brothers and herself, so strong and infectious was this sensation of fear.

So that the power of sight at this time would only have added to her curiosity and in no way have helped her understand what was happening around her, for it was too dark in the lair. Sight would only have impelled her to satisfy her curiosity by attempting to see what, for instance, lay beyond the dim glimmer of light at the entrance of the lair. Her mounting curiosity would have led Scrap to leave her mother's side almost inevitably to find her way out of the lair, and become immediately vulnerable to the dangers outside.

Born blind, she was from the start obliged to learn to use her powers of scent, hearing and touch, which would be most useful to her in the darkness. From the start she was taught not to rely on sight at all, and it wouldn't be until she had really absorbed all the information she could pick up through her nose and ears, that her eyes would open. The ability to see, the power of vision, was what you might describe as a kind of bonus. Though that world you were to see was never to be much more than a world of shadowy shapes, a world of monochrome, without the bright blues of the sky, the golden gleam of sunshine or the green of the trees, the silver of the birch-bark. None of this would Scrap ever know. She would always be colour-blind.

Now her eyes were beginning to focus themselves on her shadowy brothers, while they themselves tried to focus their bright blue eyes on her and each other. Her mother was out hunting. Scrap squeaked in the darkness, longing for her return. She and her brothers mewed and played games with the weakling inevitably getting the worst of it, and then suddenly Scrap found herself tense. She was receiving information, a scent which she knew instinctively meant fear. It wasn't the scent of her returning mother nearing the entrance to the lair. It was quite a different scent, a scent that she had never known before. It was the scent of man.

Chapter Four

The wildcat was heading back to her kits. She was just emerging from the pine-trees when she paused to lick the tip of her nose with her tongue – the same as if you want to learn the way the wind is blowing and there is only a slight puff or two, you lick your finger, hold it up and the side that feels cool shows the direction from which the wind is coming. The breeze was light and the wildcat was very jumpy. Which was why to make sure she knew where the wind was coming from she licked the tip of her nose; she didn't want to miss any tiny item of information that might be on the air.

When she had left the lair half-an-hour earlier, her kits huddled together for warmth, whimpering and squeaking for her, she had felt a tremor of apprehension. There had been something on the wind which foreboded danger. She had waited several moments but there was no further hint of warning, though she had remained anxious, still tense, and she had decided to continue with her hunting. Now, on her way home, pausing with the resinous scent of the pines heavy about her, like the smell of the distant sea, she could still pick out another scent which troubled her.

However slight a breeze, in the direction from which it blows must lie the source of the scent it carries and its precise identity. Suddenly the wildcat stiffened. She moved upwind – that is, against the direction the wind was blowing. Constantly as she went forward she moved her head from side to side, her nose quivering all the time. Sooner or later she would pick up the signal she knew instinctively was in the air. She stopped, head lifted, ears swivelled forward. She moved a little to the right, then to the left, testing the wind all the time. Suddenly she began to slink close to the ground.

She caught the scent of human beings.

Fear struck her so that every muscle became rigid; she froze like a statue, taut and tense, her jaws wide, her eyes slits once more. Then she relaxed a little and began to move forward, keeping to the edge of the trees. She gained the top of the small incline from which she could see her lair. She made out the figures outside its entrance. She heard their excited voices.

For several seconds she stood there, terror striking her on the one hand and, on the other, horrified anxiety for her four kits. For a few seconds the two emotions fought for ascendancy, but in the end fear of the human beings won. Turning aside she gave one backward glance at the figures whose excited voices reached her. Then she turned and ran back the way she had come. Back into the shelter of the forest.

She did not go very far. The pull of her kits began to turn her, and her head swung leftwards.

She began to make her way back to her lair, not taking a direct path but swinging left in a semi-circle. She would come down to the vicinity of her home from the part of the forest almost directly above the battered oak-tree. She moved with silent speed, pausing every few yards to sniff the air and wind. It was not only the human beings that she was still able to scent but also other animals and birds. Somewhere to her right came the whiff of smell from a badger's set. Now the scent of a rabbit touched her nostrils, but she made no pause, moving on swiftly. The wind rustled in the branches of the birches and the larches above her.

As she neared her lair, the smell of human beings was not so strong. They had moved away. She quickened her pace. Now she gained the undergrowth that bordered the hillside which led down to the entrance to her lair. She slid through the undergrowth, then paused at a point of vantage from which she could look down. She heard the sound of voices. Now she could scent the human beings more strongly. She saw figures receding into the distance and knew that the human beings were going away. But she also knew without any doubt that the hiding-place of her kits had been discovered.

She waited for several minutes until certain that she wasn't risking being winded by the human beings, who by now had vanished from sight. She sped down, still keeping a semi-circle until she reached the entrance to the lair. She turned to sniff the air again, reassuring herself that she was unobserved and disappeared between the roots of the oak-tree. Scrap and her brothers squeaked at her. The elder brother roughly pushed Scrap away as she sought her mother's teats. But her mother didn't give Scrap or her brother or any of them a chance to feed. She grabbed the weakest of the kits by the nape of its neck, hurried out of the lair and, still taking a circuitous route, came a few minutes later to the entrance to a rabbit's burrow. She went in, the kit between her teeth.

The burrow had in fact been half-finished. Her mate had killed the occupants, a pair of adult rabbits with their new-born youngsters, earlier that spring. There was only the one entrance to it, it wasn't a warren with half-a-dozen entrances and exits; just this single burrow which its previous occupants had scooped out and in which they had made a nest for their offspring.

Quickly the wildcat dumped the kit, and, while it mewed and squeaked, was already out of the burrow and heading back still in that roundabout fashion to her old home. Scrap was next; she too was whisked off and dumped unceremoniously beside her brother. She squeaked a little, and her brother mewed at her as if to ask what was going on. But Scrap remained beside him. Naturally, she was filled with burning curiosity – it isn't every day you're whipped away from a warm, comfortable home which you were just beginning to explore, and dumped in another place, almost without realising it has happened, and yet with this feeling of danger in the air. And so she remained where she was in the secure darkness of the burrow, comforting her brother.

Twice more the wildcat made the journey, each time carefully sniffing the air as she left her old lair and all the time as she made her way, with a kit clutched by the nape of its neck between her teeth, back to the burrow.

All the time she was filled with tension and apprehension which communicated itself to her kits so that each one squeaked and mewed during the journey. But she made each journey safely and at last was able to circle her body round Scrap and her brothers on the warm earthen floor of their new home. As she suckled them she relaxed, then the tension that had gripped her the past half-hour slipped away. She even purred a little towards her kits and they relaxed and Scrap's tiny tail flicked with pleasure and a sense of security.

Finally Scrap and the three other tiny shapes lay together replete with their mother's milk. One after another they fell asleep, although all the time the mother's ears were alert for the slightest sound. While she knew that they were safe she also knew that it might not be permanent. At any time the human beings might discover her new lair and she must be prepared to find yet another home if necessary for herself and her kits. As she lay there drowsily waiting for sleep to overtake her she sniffed the scents from outside the burrow, scents on the wind from the loch. There was no longer the smell of the old oak-tree; the earth of the burrow was a new smell. But there was no smell of man.

Her kits mewed sleepily. Scrap squeaked as her strong brother shoved against her in order to lie more comfortably. Scrap shoved him back, and her mother licked her tenderly. And then she, too felt her eyelids close drowsily, and she herself fell asleep.

Chapter Five

Hamish Munro finished the tea which his wife had left him before she went into nearby Kingussie to do some shopping. He wiped his ginger moustache, called his dog, Heinz, and took his shepherd's crook and went out of the front door. He was a Badenoch Forest gamekeeper, and gamekeeper to the same laird as had been his father and his grandfather before him, just as his cottage was the same cottage in which his father and his grandfather had lived in their time – though now it had been fully modernised. Outside there was a garage for the car. All the amenities that a head gamekeeper might expect.

Though gamekeeping was not what it was. Once the top job in the Highlands, the gamekeeper himself ranked above gardeners, farm-workers, grooms, and he would have had half-a-dozen under-gamekeepers. Now he had only one assistant. He could remember the time when his father was almost a confidante of the laird. When he had been assistant to his father in those days, from June until October special trainloads brought the guests and their families, with their valets and their maids, come up to the Highlands for the shooting, the deer-stalking and the fishing. Guests came for a fortnight at a time and when they came they welcomed the gamekeeper as if he were an old friend, and when they went they left fat tips.

And Hamish Munro's grandfather had been a gamekeeper

when it was the great age of gamekeeping, before the decline in game preservation in the days when the guests were personal friends of the laird. Between the wars the change had begun. Now they were mostly paying-guests, handing over large sums for the privilege of shooting, stalking or fishing over the laird's land.

All the same, although the golden days were gone for good, Munro managed to retain a deep pride in his job. He still felt himself of the elite, his word was law on the estate, he still wore well-cut tweed breeches and jacket.

He had been away to the war and it hadn't been a good war for him, he had been wounded in his left hip, and on his return to Civvy Street had endured plenty of trouble with his hip, necessitating several operations which had left him with continual "rheumatic twinges", as he preferred to describe them. He was always very touchy about having been wounded, as if he felt it had been all his fault, that he ought to have kept out of trouble and stayed in the war.

The shooting wasn't so extensive now, it was proving too expensive for the laird to keep up as a hobby for himself. He kept it going only for a few weeks during the summer for the paying-guests, Americans and people from London. So the tips and the perks weren't so fat and they were less frequent.

Limping slightly, Munro made his way along the short path to the gate and, closing it behind him, headed in the direction of the forest. At his heels was his mongrel terrier, Heinz. Heinz was running to fat a bit. The dog rushed off ahead and then turned back to Munro, his pop-eyes shining, and then together they set off towards the edge of the forest where it joined a long stretch of moorland, and half-a-mile ahead round the corner where the forest sprawled out on to the moorland lived a dozen caravanners; they had been coming here the past three years.

It was a fine late afternoon and this sort of mild weather was good for his hip. Once he got moving his limp hardly showed at all. The air was clear and fresh. He could smell the scent of the waters of Loch Insh on the wind. A sense of well-being flooded him and he paused and took out his pipe and tobacco-pouch and filled up for a smoke. Heinz ran on ahead for a few yards and then turned and looked towards him and came trotting back.

Mungo lit his blackened old pipe with a match – he didn't hold with cigarette-lighters; you could taste the petrol when you smoked – and tamped the tobacco down and exhaled a cloud of smoke. He drew at his pipe a couple of times and, satisfied it was going well, he set off again along the edge of the forest. He had in mind to look out for the traps that he and his assistant had set earlier in the day to catch foxes or wildcat.

There was always the possibility that someone, a caravanner or some stranger venturing into the forest, might disturb a trap. He wanted to make sure this hadn't happened.

The forest was late in leaf, the trunks and branches of the oaks and the silver birch, the spruces and the pines and the aspens, were all twisted and gnarled with thick crumbling lichen as a result of storm and blizzard, rain and frost and snow which had battered the forest late into the spring. You wouldn't recognise the gnarled, almost sinister-looking silver birches, for example, especially in the twilight shadows, compared with the slender silver birches on the way down to Inverness itself, where the fields and woods seemed to be much greener and sunnier than they were in this part of Inverness-shire.

As he and his dog rounded a fringe of the forest that jutted out onto the moorland there were some caravans to his left, and then at the same time he heard voices coming from the forest itself. He paused for a moment. He caught a note in the loud voices which made him wonder what was going on.

As he got nearer he saw four men two or three feet up on a bank over which hung the branches of birch-trees. He saw the flash of a spade. They were digging out the earth and he remembered that this was the spot where there were some rabbit-burrows. Munro took his pipe out of his mouth and pulled at his moustache. He thought that the men had probably found a badger, or a fox holed up. He went forward quickly. At the same time Heinz rushed ahead barking aggressively.

The men turned to Munro, the one with the spade stopped digging. "It's a wildcat," somebody called out. "This is her nest," someone else shouted. The first man said that he had seen a wildcat carry a youngster into the burrow earlier on. He'd caught sight of her as she'd disappeared. He went on about having seen the same wildcat – he reckoned it was the same one – going into a cave under an old oak further in the forest.

Hamish Munro realised that the man was probably right. There were wildcat in the area, he knew, because he had found traces of their hunting activities. He had passed the rabbit-burrows many times, of course, but it had never occurred to him that it might turn out to be a wildcat's lair. Though he remembered that yesterday Heinz, when he had passed that same place, had growled and his hair had bristled a bit. Obviously he had been very interested, but Munro had called him off because he thought that it was only a rabbit that Heinz had sniffed up there, and there were more important things that he was after. Such as wildcat and foxes.

"It is none of your business what's in there," he was saying to the men "You're trespassing – you've no right on this land at all."

The men muttered apologetically and began to withdraw slowly in the direction of the caravans. The spade glinted in the sunlight.

"We were only going to destroy the mother and her young," the second man said as he went past. "I reckon she's there with them."

Heinz barked at them until the quartette were gone. Munro called the dog to him, clamped his teeth on his pipe-stem, and went to the rabbit-burrow where the men had been digging.

Chapter Six

He took his pipe out of his mouth, and tugged at his moustache reflectively. He was recalling last January, late one afternoon, when he'd been out and seen footmarks in the snow which lay heavy on the ground. They were those of a wildcat. He had tracked the footprints through the forest and out onto the moor. He had come to a sudden stop where the footmarks ended suddenly. He had gone on several paces and there he saw blood and partridge-feathers on the snow. It was obvious what had happened. The wildcat had pounced on an unsuspecting partridge, jumping a clear eight feet, which was why the footsteps had stopped.

Then the wildcat's footprints had continued, together with signs of the partridge's feathers and drops of blood. Munro had followed and then had seen the wildcat itself, the dead partridge in its jaws. Realising it was being followed, it had dropped the partridge and headed down towards Loch Insh. It ran along the shore for about fifty yards.

He had been by chance without his gun, but nevertheless had raced after the wildcat. He saw it turn and dive into the loch and swim for over five hundred yards before it turned back to the shore again. It made off along the shore leaving Munro to watch it until it disappeared.

The wildcat spends a lot of time in trees. It is really an animal of the forest. It's dark stripes against its tawny background mix in with the colour of the branches and the foliage. The wildcats of that part of Inverness-shire were supposed to prefer larch trees, Munro had always heard. There was this predilection amongst wildcats in the local forests to get up into a larch tree. Next to that they liked to use the spruce. The way a wildcat hunts is by surveillance and stealthy stalking of its prey. That was one reason why it climbed a tree from which it could watch over quite a considerable area, keeping an eye open for the movements of voles, rabbits or small birds prior to pouncing on them for the kill.

Driven out of the forest by man hunting it and also by clearing for agriculture and the stripping of the forests of wood for building purposes, the only place left for the wildcat were the rocky hill wastes. It had acclimatised itself to this new environment, making its den in rocky caves, and would lie on the rocks to sun itself on warm days.

Its colouring – its camouflage, if you like – was equally suitable for its new habitat, its sandy, tawny background to its stripes blending with the rocks, sand and heather and the juniper trees. Munro had once sighted through his fieldglasses a wildcat lying on some rocks in the summer warmth; and he remembered coming across something like a hare's form in a forest clearing among fallen fir-cones. It had been where a wildcat had been sunning itself and had left behind some of its hairs among the cones.

Now he called Heinz. With only a word from him the terrier was sniffing at the rabbit-burrow. At the same time he could hear mewling and squeaks down there in the darkness. In a flash Heinz went down and only a few moments later he was back with a wildcat kit in his jaws, its neck already broken where he had snapped it.

No sign of the wildcat mother and Munro guessed that she was waiting for the human beings to depart so that she would get the chance to remove her kits and hide them somewhere else. There was no sign of the male wildcat either. But he knew that, even if he was somewhere in the vicinity, he would have had nothing to do with the kits after they were born and being brought up by his mate.

As Heinz brought out the dead kit, looking no larger than a mouse, its head on one side where its neck was broken and laid it on the ground for his approval, Munro spoke quietly to the dog to hold back. He took the shepherd's crook and pushed it down into the burrow. He felt the kits at the other end and could hear their squeaks and whimpers, louder now. He hooked out one and although Heinz had a go at it, even inflicting a nip on it, he pushed him aside and shoved the kit into his capacious jacket-pocket. Heinz watched him with staring eyes, growling disappointedly at not being able to finish the kit off.

Munro ordered him to keep out of it while he reached down again with the hooked stick and pulled out two other kits one after another. They still looked no larger than mice and their their bright blue eyes couldn't focus and they were dazzled by the brightness of the day. He calculated that the kits must be just over ten days old and that their eyes had barely opened. Before he put the fourth kit in his pocket, he looked at it. Heinz jumped up at his hand as if trying to grab it, but Munro cuffed him down and Heinz thereupon dived down into the lair once more. But this time all that could be heard were his growls and scratching sounds. There was no whimpering or squeaking, and he knew there were no more kits down there. He was, in fact, pretty sure that four was the most that a wildcat ever had of kits. He stared down at the tiny mouse-like shape in the palm of his hand. He noted the tiny stump of tail and the miniscule paws extended. The little body was tawny-coloured, with the typical tiger stripes not yet black but a kind of greenish brown.

As he put it into his pocket, he paused. Ordinarily he would have let Heinz go in and finish the job and be glad to see him kill the wildcat's offspring. What had made him relent for the first time in his life, he who had killed plenty of wildcats? The presence of the caravanners trespassing, with their shouts that the wildcat and her kits ought to have been killed? It was no job of their to kill them; had he resented their interference? Had his antagonism towards them for once put him on the side of the kits?

Heinz reappeared from the rabbit-burrow growling in disappointment and as if to make up for his feeling of frustration he worried the little kit he had killed. Munro called the dog off, and turned back for home, Heinz running ahead of him and then running back to jump up, knowing full well he had those three kits in his pocket. He could feel them warm against his "rheumatic" hip. And he remembered what it was had urged him suddenly to feel protective to the three tiny creatures in his jacket-pocket.

After a few yards he turned and looked back. Had the mother wildcat shown up? But there was no sign of her, and he turned and went on his way.

Chapter Seven

Scrap had heard the voices outside and simultaneously scented the human beings, as had her three brothers – although the weakest of them had been a bit slow on the uptake as usual. Instinctively she had sensed danger. She at once stopped playing rough and tumble games with her brothers and stiffened. Apprehension filled the dark burrow.

Scrap turned as if to find her mother, but at once realised her absence and instead turned to the strongest of her brothers. But he himself was concentrating too much on what he could hear and what he could smell from outside the lair to take any notice of her. He spat and snarled and crouched, his tiny tail held up straight behind him.

This was the dark place where she had been dumped the previous day, carried there hurriedly, glimpsing the strange brilliance of the sky, her eyes dazzled by the light of the day; and all that fear and tension vibrating from her mother, and the confusion of new scents of the forest and of the earth in this new home. There had been the security of the darkness once again, the relaxing of tension, the warmth of her mother; her brothers' proximity, and their settling down after the squeaks and grumblings over being disturbed so suddenly had ended. And there hadn't been any more scent of human beings.

Until now. The scents, once more, smells of men, and voices, filling the air with fear.

Scrap crouched there waiting to learn what would happen next. The voices continued; there were more of them, it seemed. There were more smells, smells of human beings. Smells that Scrap and her brothers feared instinctively more than any other. Scrap caught a clink of steel. Striking against rock. She could smell the steel. The scent of it began to mingle with that of the others in the burrow. For the next several minutes Scrap waited with her brothers. She imitated her strong brother, as did the others, even the weak one, snarling and spitting, longing for the return of her mother so that she could shelter against the warmth of her strong body.

Now Scrap heard another voice, she caught another scent. It was a man's scent similar to the others, but there was something else added to it. The smell of violence. Violence that threatened her and her brothers and was dreadfully close at hand. And with this new scent came another one, that of some animal. It was strong, pungent, and that, too, added to the scent of violence that was now crowding into the burrow. The clink of steel against rock stopped. The sound of the voices grew louder and now there was one louder voice. Then very soon the other voices seemed to fade away. The scent of steel vanished. The scent of the other human being was still there.

And there remained this other scent which assailed Scrap's nerves and every fibre of her tiny body. All about her now the scent of fear was building up. She knew that her brothers, even the strong one, were filled with apprehension. Her strongest brother snarled and spat, it now seemed automatically. She and the others followed his example.

Suddenly that smell of the animal was nearer. Scrap could hear a horrific growling, the noise filled the burrow, and she and the stronger brother backed against the end of the burrow. At the same time the light at the entrance was blocked out by a snarling, growling shape. The second brother followed suit and the three of them were backing against the back of the lair leaving the weaker one to look after itself.

It was the weakest of the kits that was the first victim to this horrible, snarling shape that pounced upon the tiny body.

Scrap could smell the saliva that dribbled from the animal's jaws. Its growling and snarling filled the burrow. Scrap could see the flash of the animal's white teeth close over the neck of her third brother. She heard him give a tiny broken screech, then there was a click of snapping brittle bone and the animal was backing out of the lair, dragging the tiny kit after it, growling triumphantly.

There followed a few moments' respite during which Scrap heard the human being's voice raised, and the animal's barks and growls of satisfaction. Her head was filled with a confusion of images. That terrifying shape; and once again she saw the flash of its teeth, huge and ferocious, the frightening glare of its white eyeballs. She smelt that overpowering smell.

But now something else appeared at the entrance to the burrow, blocking out the light. A hook reached down, probing, inescapable. Scrap spat and snarled and clawed with her tiny paws at this new terror. It was no use. However, her resistance, such as it was, and that of her stronger brother helped put off the menace for a few seconds. It hooked the other kit. The kit screeched and squirmed in its implacable grip, but there was no letting go. Now the hook was jerked back, dragging the kit with it.

Again the looming shadow at the burrow-entrance, and the hook reached down. It curled round her brother's body and jerked it, clawing, hissing and snarling, clean out of the burrow. There were the animal's growls outside and yaps of excitement. The yaps ceased as the voice sounded again.

Scrap was now alone. The hook came down again and although she tried to avoid it, it curled round her little body. Much as she fought and clawed, it was of no avail. Then a few seconds of bright sunshine. Again she caught a glimpse of brilliance overhead, glimpsed the forest round her, the tall trees with their branches beginning to burst with green.

Then the hand that smelled of human beings grabbed her and unhooked her from the shepherd's crook and thrust her into a pocket; and she was again with her elder brother and the other one. There was squeaking and sniffing as they felt themselves being jogged along in the total darkness of strong-smelling tweed and the odour of old tobacco. More new scents for Scrap to sniff, new scents to add to the collection she was storing up in her memory.

As they continued their jogging progress, Scrap and her brothers began to relax. Other scents reached her through the thick material of Munro's pocket, as well as the smell of the man and animal there were the scents from the road, the earth, and other scents borne on the wind from the loch. There was an occasional whimper of pain from the kit the dog had nipped, and Scrap comforted him and felt the abrasion on his leg. She began to lick it gently with her tiny tongue in an effort to heal the wound.

The jogging relaxed her and she began to doze. She and the two others snuggled up to each other, adapting themselves to the movement of the pocket against the leg of the human being who had removed them from their home.

It was quiet now where the edge of the forest met the moor. The sky was beginning to turn a speckled grey of dusk. A bat dipped and swerved, feeding off the evening insects. Suddenly a shape stood at the top of the bank in the shadow of the birches. It was the wildcat mother.

She paused irresolutely for a few moments sniffing the air. Then very slowly, head shifting from side to side, she came down, making a circle, but heading all the time in the direction of the rabbit-burrow. A few feet from it she stopped. Sniffing more decisively, she hurried to where the tiny inert shape of the dead kit still lay on the earth which had been disturbed by the feet of human beings and the paws of a dog.

She stood over the dead kit, then bent her head and began

licking it slowly and steadily, as if she could restore it to life. Suddenly she raised her head and stared in the direction of the forest as the male wildcat made his appearance.

He too paused for a moment, looking in her direction and sniffing the air for the scent of danger. He winded her and came towards her. Several yards away he stopped again and the wildcat mother drew back the striped mask of her face in a snarl, hissing her signal that he should leave her with her kit. Though, all the time she had been rearing her kits, she had warned him to keep his distance, for they evoked no paternal love from him – they represented nothing more than potential rivals for his mate and his territory – he would not hesitate even now to devour this dead offspring.

The wildcat mother picked up the kit tenderly in her jaws. A forlorn, drooping figure, she moved slowly to the entrance of the burrow and then disappeared inside. The male wildcat stared after her hesitating. He made as if to follow her and then stood still again. For a few moments he sniffed the air about him, eyeing the bat as it dipped and swerved a few feet above him. Then he turned back into the trees. Some twenty yards into the forest he paused at the foot of a larch tree and sprang up to disappear in its branches.

Back at the burrow the wildcat mother reappeared, without the kit. She had left it, earth scratched over it, down there in the tomb that the burrow had now become. She paused at the entrance for several moments, head moving restlessly from side to side, and then raised her head and let out a screeching howl.

A cry of sorrow for her lost children? A cry of mourning for the one she had left behind her, dead? She howled again, a long drawn out miaow which reached its harsh climax in an agonised screech. From the forest a tawny owl replied. It was itself just setting out to hunt for food.

And then came the answer to the female wildcat's cry. A harsh scream from her mate high up in the larch-tree. The mother wildcat cried again but this time less loudly. There was a sound of finality about that last cry, and about her forlorn shape as she moved off slowly towards the forest there was a droop of resignation.

Again the cry from her mate in the forest. And then she herself had vanished to find him to seek comfort with him for her loss.

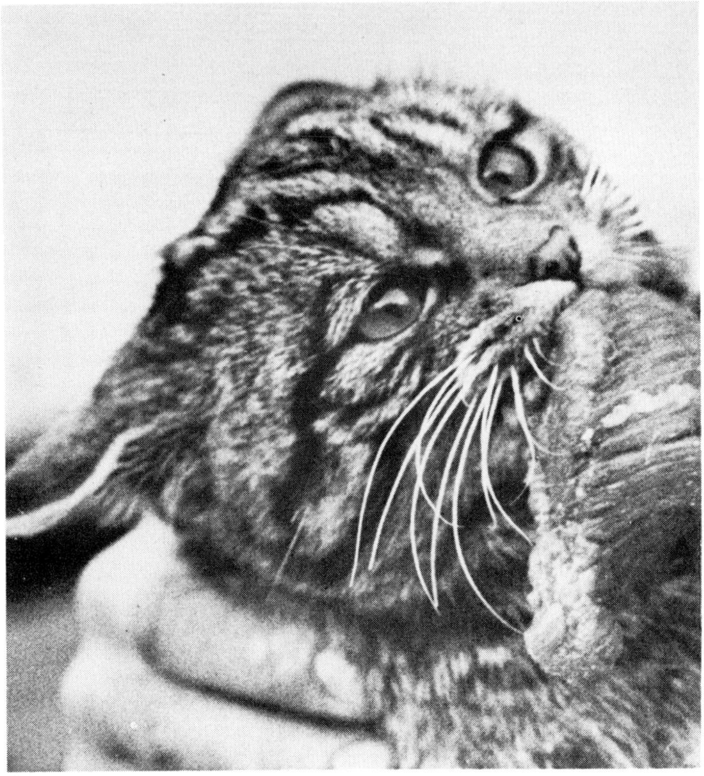

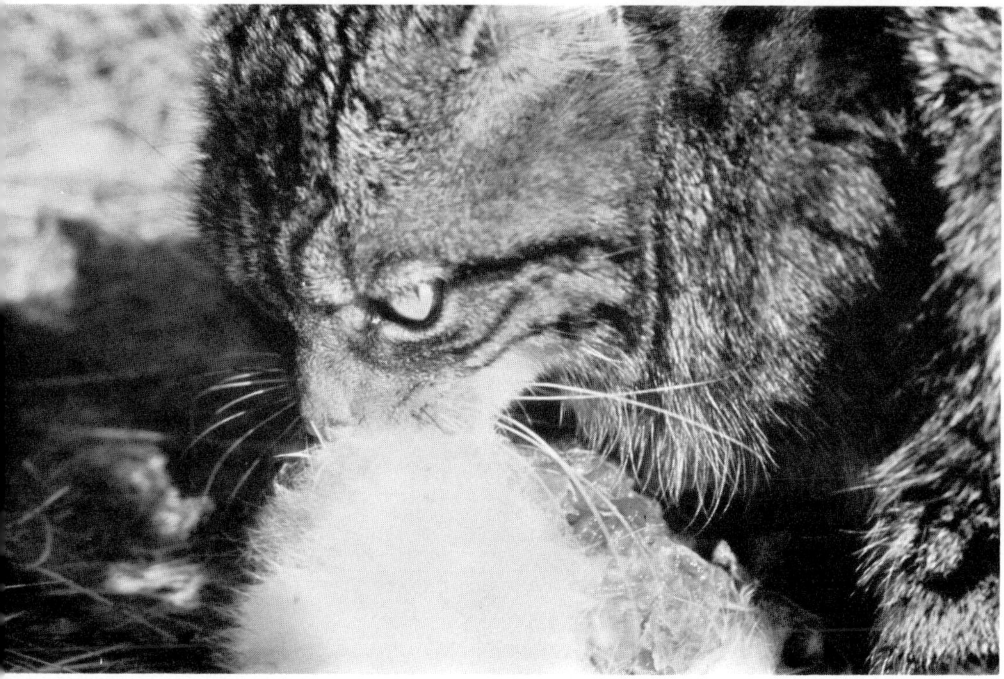

Scrap with the author. This photograph is almost certainly the first ever taken of an adult wildcat with a human being

Scrap, as a kitten, plays wi Colin Burton as the childr look on

Sanders

Sanders and Scrap
vith Muriel Burton

PART TWO

Chapter Eight

When Hamish Munro came out of his cottage again, dusk was falling. His Belgian-made double-barrelled 12-bore was tucked under his arm. Cost him a tenner before the war. Worth at least fifty today. If you wanted to sell it, that is. Which you never would, not this one. Heinz trotted at his heels, every now and again running ahead and waiting, eyes popping up at him, expectantly.

He had left one of the kits with his son, Donald, who had hopes that he might rear it; he had taken the one that had been nipped by Heinz. Munro foresaw that it wouldn't recover from the shock of the nip and consequently wouldn't last long enough to cause any trouble, anyway. The other two, Scrap and her brother, he had left with Mr Orbell, the manager of the Highland Wildlife Park at Kincraig, a couple of miles away. It was he who had asked him to keep a look-out for any wildcat – alive – and he'd be glad to take them. This must have been what had stopped him from letting Heinz kill the rest of the kits earlier on.

The object of the park was to bring back to the Highlands animals that had once lived there – the wild boar, the bear, the wolf – which had at one time freely roamed the forests and glens before being wiped out by human beings. The wildcat had escaped extinction by the skin of its teeth. As it was, only a few remained, and those tried to stay out of sight as best they could, keeping to the loneliest rocky hills, sometimes venturing into the forests.

Catching a fully-grown wildcat, alive, wasn't as easy as falling off a log, exactly. You needed to trap them in such a way that they would be unharmed, and you needed thick gloves and thick clothes as protection from their slashing claws which could tear you to pieces – much easier to shoot them or trap them in the horrific gin-trap. But obtaining wildcat kits whose parents you had eliminated was another matter.

He had said to his wife that he thought perhaps one of the caravanners, disappointed at not being able to get at the wildcat kits, might think that it was clever to see if he could find other wildcats' lairs in the forest. He would have no idea, Munro reminded her, being a stranger to the forest, of what danger he could run into if he did meet a wildcat face-to-face. Slash you to shreds a wildcat could, especially if it was in a tight spot, and you tried to capture it.

He caught the cry – *coor-li . . . coor-li . . .* – of curlews from the marshy shores of the loch, that strange, eerie note that always made him cock his ear, though he heard if often enough. But it was a haunting cry, something almost human about it, as if from some wandering soul out there as the dusk was falling. He made his way along the road towards where the edge of the forest protruded onto the moor. It would be a new moon and no stars had yet appeared in the sky. The bats were beginning to fly low over the road. From the forest he heard the call of a tawny owl. *Kee-wick . . . kee-wick . . . kee-wick,* the familiar cry reached him and he heard the owl's mate answering in the distance. *Kee-wick . . . kee-wick.*

He took his pipe out of his mouth, and pulled at his moustache. He tensed; he was near the spot where he had come upon the caravanners attempting to dig out the kits earlier on. He paused and sniffed the wind, while Heinz, whining a little, looked up at him impatiently. He ordered him to keep quiet. The dog obeyed and stood close at heel.

Instead of approaching the wildcat's lair as he had done earlier on, this time, the tension in him relaxing now, he moved past it and turned up into the forest so that he would approach it without being winded by the wildcats. For that was his purpose in this second visit to this part of the forest. He had a pretty good idea that the female wildcat would return in search of her kits. A shadow through the darkening forest, he approached the place, Heinz padding beside him noiselessly. There was the earth which had been shovelled from the burrow by the caravanners lying in loose heaps. He stopped and gazed round him slowly and then glanced down at Heinz. The dog had stiffened, its nose raised and the hackles lifted along the back of its neck.

Now, every nerve in his body tingled. He knew that the female wildcat was somewhere around.

He decided she wouldn't be in the lair. She would have gone there straight away, which would have been some time ago by now, and found the kits gone. She wouldn't have hung around for a moment longer than necessary – the smell of human beings would have been too heavy on the air to be comfortable for her.

He moved forward, his gun ready. Now he was looking up at the larch-trees of which there were any number in this part of the forest. He glanced down at the dog. Heinz too was looking up into the trees.

From long experience with Munro he knew, too, of a wildcat's liking for the security of trees. He was sniffing the air with his up-tilted nose as he went forward slowly. He stopped. Munro glanced at him. Into which tree had the wildcat climbed?

Then Heinz started growling, his teeth clicking. He had found the tree.

Gun snug at his shoulder, Munro's eyes narrowed as he moved to the dog's side. No tension, now, but relaxed, almost nonchalant, he watched for the faintest movement in the branches that could betray the female wildcat's presence.

He didn't have very much time. Already the forest was darkening, the pale dusk was turning into the blue-grey evening.

Suddenly he stiffened, and he felt the dog close by him tense as well. It uttered a low growl. His trigger-finger squeezed, but he fired into the air. Still he kept his eye fixed on the spot in the branches where he had caught a movement which was not that of the breeze in the tree-tops. Almost as if it was a puppet pulled forward on a string a wildcat's head appeared round the side of the tree-trunk.

It was an old trick.

A wildcat, like most animals, is always very curious: any unusual sound is irrestible and will attract its attention. So this is what you bank on. One shot in the air to attract its curiosity and make it show itself, the next one on target.

Munro's finger squeezed again and this time the shot blew part of the female wildcat's head off and she fell to the ground. Heinz rushed towards the shape at the foot of the tree and worried it until he could get his teeth round its neck, when he started to drag the body towards Munro.

But Munro's eyes narrowed, the ginger moustache seemed to bristle above his tightened mouth. His second shot had produced an unexpected effect. Even as the female wildcat fell he had caught a movement in a larch-tree to his right. As he looked up he saw another wildcat's face appear from behind the tree-trunk. Again he fired, and this time it was the male that crashed to the ground.

As he lowered his gun and walked over to his second victim he thrilled, a surge of elation swept through him. The stiffness in his hip seemed to vanish, and he stepped out as springily as a youngster. His earlier moment of tender-heartedness towards a hated enemy had paid off, after all. Two adult wildcats killed, one kit dead and three other kits prisoners-of-war as you might say. Not a bad score. He turned the dead male wildcat over with his toe.

Calling Heinz to leave worrying the female and to come to heel, he headed off back home, gun under his arm; he was looking forward to the tale he would have to tell his wife over a dram.

Chapter Nine

Scrap was trying to assemble and memorise new scents and signals and store them away inside her tiny brain. The scents of the man and his dark, tweedy-smelling pocket, and the animal that had rushed so savagely into their new home – that smell was still around, and Scrap knew the animal was near. There had been the light in her eyes, when she and her two brothers were taken out of the darkness of the pocket, and another human being's scent. Then the attempts to feed her, which hadn't been at all like her mother suckling her. And the mewing protests of her brothers. And always the sound of human voices. Two human faces bent over her. And she was unable to focus them properly, her eyes dazzled by the bright light of day.

Then it had been a return to the dark, tweedy-smelling pocket, and now only one brother with her. The strong brother. The other, whose wound she had tried to lick to comfort him, had been left behind. Now there was another sound, the rhythmic sound that was metallic, there was the slam of metal against metal. A slightly jogging motion and the warmth which she had already experienced of the dark pocket, and she and her brother curled up together.

The rhythmic jogging movement stopped. There was the metallic slam once more and the slower jostling movement for a few moments and then she was placed on a square flat surface and again scents and human voices filled the atmosphere, again there were faces bent over her, two different faces from those that she had smelt before.

Now she and her brother were in a straw-lined box; human hands, gentler hands that those before, were helping her into her new home together with her brother. Then something was being held against her mouth. She could get her mouth round this teat all right, and she could taste the sweetish taste. Dribbles of it reached her throat and she swallowed. It tasted very much like her mother's milk and she swallowed some more. Her eyes were not focusing very well, but she managed to watch her brother while he received the same treatment.

She could see the milky fluid dribble down the side of his jaw and she reached out and patted it with her tiny paw. Human beings' voices floated over her, human beings' smells encircled her. She was given some more of the fluid. Once again the teat, though it wasn't soft like her mother's, was pushed into her mouth and she sucked down the milk as hard as she could. Finally she gave a burp, the milk was taken away and she slumped down beside her brother. She saw that his eyes were closed. As she pawed at him gently they flickered open for a moment, staring at her, unfocused, and then closed again.

For a moment or two she looked up at the faces bent over her, then she too closed her eyes and giving another burp fell into a doze. She allowed herself to drift away into sleep.

When he reached home Munro found his son trying to feed the kit with milk from an eye-dropper, but tiny as it was, the little animal was putting up an aggressive show, spitting and scratching with its needle-sharp claws.

"Ye'll never get anywhere with it," Munro told his son. "Ye'll nae tame, it, it'll never make friends with ye." He tried to help Donald handle the kit to get some milk down its throat. It turned on him and inflicted several scratches on his hands. He shrugged off any more responsibility towards the kit. Best thing, he said, was to give it a good crack over the head, finish it off there and then.

But Donald wasn't for having it destroyed. Not yet at any rate. It was then Munro had an idea. Why didn't his son, he said, give the kit to Colin Burton the head warden up at the Highland Wildlife Park? Mr Orbell had already taken two of the kits, so that would be enough for him to look after; but the park was still in need of more wildcats.

Colin Burton would be glad enough to take on this little kit if it managed to survive. Which Munro thought was highly unlikely, the nip Heinz had given it hadn't done it any good. What was the point, he said to his son, of going to all this trouble when it would die anyway; and if it doesn't die it'll nae thank ye for trying to save its life, but only turn on you? That was what was wrong with wildcats, they were totally untameable, unmanageable. This one might quieten, might behave all right for a few weeks, although even at this early stage it was showing its aggressive nature. But in the end the wildcat, even if it did seem to become tame, would inevitably end up by scratching him to ribbons.

"I'm sorry I gave it tae ye. 'Twould be best to bash it over its heid, like I said. Or, onnyway, tak it up to Colin Burton at the Wildlife Park."

Chapter Ten

Colin Burton had finished his supper on his return from the park and was drinking coffee when there came a knock on the caravan door.

At once Sanders, Colin's 16½ year old dog, named after Edgar Wallace's famous fictional character, Sanders of the River, because he loved splashing about in the burn, started to bark. Colin told him to shut up, while he went to the door and invited young Munro in. Sanders still barked. He enjoyed barking; it was his way of showing to visitors that he might be getting on in years, but there was plenty of life in an old dog yet. He was black with a very grizzled muzzle and was a cross between an English curly coat retriever mother and a working Border Collie father, and Colin had owned him since he was a schoolboy in Penzance, Cornwall, where he was born.

Finally, Sanders shut up and Munro's son told Colin about the kit, how his father had come by it. On his way home Colin had already learned that Eddie Orbell had obtained a couple of kits from Munro himself earlier that evening. He took the tiny creature in his hand while Sanders began to bark his interest in the newcomer, and Colin had to turn away to silence him; he limped away into the corner of the caravan. He not only creaked a bit with old age and arthritis, but he had a permanently injured right foreleg.

Donald Munro explained how he hadn't had much success

getting milk into the kit with his eye-dropper; and Colin had at once noted the wound it had suffered from Heinz's nip. Privately, he didn't give the kit much hope of survival. Oddly enough the kit was quieter with Colin, who had also noted the scratches it had given young Munro. Colin decided that in fact the kit had received so little nourishment it was almost too weak to react to him with the same show of ferocity that it had shown earlier to the Munros. He knew that he would have to get some food into it quickly if he was to save its life.

He invited Donald Munro to stay and have a mug of coffee, but the other didn't accept, he wanted to get back home to tell his father that Colin Burton would take on the job of attempting to rear the kit. Colin, grateful to Munro for acting with such kind-heartedness, saving the kits, sent him a message of thanks by his son for thinking of him, and young Munro left to the inevitable accompaniment of Sanders barking at him as he opened the door and went out into the darkness.

As the boy's footsteps crunched away on the gravel outside the caravan, Sanders stopped barking to listen to the burble of the burn down the bank below. Colin saw the expression on his face and quickly grabbed him by the collar, closing the door, while he pushed him back into his corner. "Oh no you don't," he said, "you're not going for another swim."

Earlier that evening Sanders had gone off as he so often did to swim and disport himself in the burn, to come back wringing wet and exhausted with chasing a bird, or it may have been a rabbit – or even an elusive fish. Now, he gave Colin his usual tortured look and then subsided in his corner. Meanwhile Colin had rummaged amongst his veterinary equipment and found a hypodermic syringe, from which he removed the needle. While he prepared some tinned milk diluted half and half with warm water he sterilised the syringe in hot water. Now he half-filled it with the warm milk and water. He held the little kit firmly while he pushed some of the mixture into it – the end of the hypodermic was small enough to fit into the kit's mouth, and at once it began suckling and getting some of the stuff down.

Colin knew that it meant he was going to be up every two hours to feed the kit. And this for the next week or more. Oh, well, he was used to it. You didn't require an alarm-clock to wake you. You woke automatically, just as the kit did.

For the next feed he gave the kit two hours later Colin added to the half-an-half measure of tinned milk and warm water a measure of Abidec, a patent vitamin supplement. He had used it before in similar emergencies and it had in fact become almost a standard for feeding young wild creatures. Every two hours throughout the night he kept the kit going with the mixture from the hypodermic.

Next morning, Muriel Bishop arrived. She was staying in Aviemore; she and Colin were waiting to get married, as soon as he had a free moment to take her over to the Registry Office, Kingussie. Now, with this baby wildcat on his hands, he would have even less time. Muriel was fascinated by the kit's minute ears, still folded as they had been at birth, its eyes still slitted, and its mouth, which kept opening like a wee bird in its nest for food.

Muriel knew that Colin was used to caring for small animals. When she had first met him at Chessington Zoo he had had four baby foxes to rear; and there had been an abandoned baby wallaby, for which a special electric-blanket pouch had been made; and there'd been a white ferret, and two more ferrets in his flat, and a baby Capuchine monkey.

Colin admitted to her that he wouldn't be surprised to find the kit had died by the time he came home for lunch. Muriel guessed that he would be back to look at the kit before lunch-time, that he wouldn't bother about lunch at all, but would make do with a cigarette though he had given up smoking two years before. She could tell by his face that he wasn't going to lose the kit if he could possibly help it.

He took a good look at the kit before he went off to the park. So far as you could tell – well, you could say it looked as if it might survive. But only might – no more than that. He was shaking his head as he left the caravan – Muriel would see that the kit was kept warm in the box provided for it, while she prepared some sandwiches and a flask of coffee against Colin's return – she herself had to go back to Aviemore shortly.

Munro's son had explained to Colin how the kit had come to receive the wound from Heinz, and although it didn't seem to have had much ill-effect, he knew that there was quite a good probability that it was suffering from delayed shock which might easily cause it to die within the next few hours.

When he got back, however, he found the kit still alive and if it wasn't exactly very perky at least it hadn't given up. He gave it some more of the mixture before he went off again to work.

He returned more to see how the kit was progressing than eat the sandwiches and to drink the coffee Muriel had left him. Well, anyway, the kit was still alive.

For the next five days and nights he managed to keep the kit going. It didn't improve, it didn't get any worse. Sometimes he thought it might be turning the corner, sometimes he felt convinced it wouldn't last the next hour.

He wasn't sleeping much, he wasn't eating much. He was back on cigarettes as Muriel had known he would be.

Then, on the sixth evening, came a knock on the door which started Sanders barking. Colin quietened him and went to the door. It was Eddie Orbell. He'd brought Scrap and her brother for Colin to look after.

Chapter Eleven

Sanders' barking when Eddie Orbell arrived at the caravan, although it had rattled Scrap's eardrums, had not filled her with apprehension as had the snarling growl of the creature that had attacked her and her brothers in the darkness of the lair. Her reaction to Sanders' bark was quite different.

But she could feel that her brother wasn't reacting as she was. He for his part had stiffened with fear, and so long as the barking continued so did he whimper. Scrap patted him with her paw to comfort him and even licked the side of his face with her rough little tongue.

Of course, there was the total difference between the scent of Sanders and that of the other animal. To start with Sanders was four times the size of the other animal that had attacked them, so there was more of him to smell. At the same time, where this smaller creature had given off a scent full of aggression, signalling immediate annihilation, Sanders' scent was again quite different. Scrap's tiny nose was able to differentiate between the two scents, the one which had caused her to fear for her very existence, and this other which told her she was among friends.

She herself was giving off a different scent from that which she had given off in the lair to the smaller animal rushing in to destroy her and her brothers. That had been the scent of fear. But while Sanders barked away she felt quite relaxed, free from tension – and the scent that she was giving off smelt accordingly. These were the sort of signals she and Sanders were receiving and giving to each other. This was how they were communicating with each other. And although they were not speaking the same language – the language of wildcat is different from the language of dog though nevertheless relying on strong powers of scent – and if Scrap's powers, she being such a tiny kit, weren't as marvellously developed as they would be when she was fully-grown, well, perhaps old Sanders' sense of smell had diminished somewhat due to age. So, you could say it was about even between them. The main thing was they were getting the message across to each other. They were able to reassure one another that they meant no harm to each other, that there was no danger in the air.

Tiny as Scrap's nose was, its sensitivity was much greater than that of an adult human being. One purpose of it was the same as that of a human nose, to warm the air as she breathed it in on its way into her lungs. The front of her skull above her nose and eyes contained the sinuses, cavities which on the one hand lightened the skull's weight while at the same time strengthened it against blows to her face and head. It especially protected her eyes. In human beings there is also the nasal duct which carries off the tears from the eyes (which explains why when you cry a lot you need to blow your nose often with a handkerchief to clear the tears away) but Scrap didn't need this nasal duct since she didn't weep tears as human beings do.

From the moment of her birth, when she'd had no eyes to see with, she had relied on her power of scent, and less on her hearing ability. From the start her nose had been continuously picking up extremely complicated patterns of scents. And right from the start she had possessed this miraculous ability to select and identify from this increasingly large number of smells those which were of importance to her at that particular moment. Like the scent of her mother when she was close to her, of the earth of the lair in which she lay, and the scent of her three brothers beside her. They had each had their own particular scent and she knew which one she was lying next to, and the exact moment when they changed position, simply by being able to memorise the smell of each particular brother.

She had known when her mother left the lair and she had known two or three moments before that she was coming back, she could scent her outside. And when her mother had arrived and brought with her the scents of the outside world: the smell of the earth she had picked up on her paws, the smell of the dew from the grass, of the leaves that she had stepped on and the scent of flowers and bracken she had brushed against and which had clung to her fur. The smell, too, of the food she had eaten and which still clung to her jaws.

One or twice also Scrap had caught another scent, similar to that of her brothers and her mother. And at once her mother's reaction, the hiss and snarl and spitting and threat to this smell which had lingered briefly outside the entrance to the lair. And Scrap's nerves and blood had responded to it in a way she had never learned to understand – and never would: the scent of her father.

As well as her powers of scent, and of hearing – which at this early age was less well developed – her sensory powers were developing fast. She was learning by her sense of touch. Touch against her fur, against her tiny, short tail; against her whiskers, those from the side of each jaw, and where her eyebrows should be. The way she had been handled, for example, when she had been picked up off the hooked stick and transferred quickly into that tweedy pocket, the roughness of his touch contrasting strongly with the gentle ways of Eddie and Joanne Orbell, when she had been given over to them together with her brother.

Their touch had been totally different from that of Munro, and although she could barely distinguish between them by the sight of her unfocused eyes, she knew at once the difference between their attitudes to her. And all the time she had been conscious of her brother's reaction to these changes in their situation, not only from the scents he gave off but by his squeaks and mews of protest and fear. She received them, all of them signals to which she had responded.

And now being transferred into other hands. The sureness of this new human being, the way he was handling her, the total lack of any apprehension in his touch. No sense of aggression here, no threat in the air, no threat to her, or her brother. No scent of fear, of violence, only warmth in the touch, and tenderness. And this is what she was trying to signal to her brother, that he was not to worry any more, that he was in the hands of love and kindness, nothing to fear now.

This was what Scrap felt when Colin Burton held her in his hands.

Chapter Twelve

Eddie Orbell and his wife, Joanne, had taken it in turns the past four days to feed Scrap and her brother from syringes containing milk, boiled water, a vitamin supplement called Abidec, and Glucodin. Later Scrap, who had settled down more easily than her brother, enjoyed a little scraped rabbit. Joanne Orbell was fascinated by the kits, especially Scrap; it was by no means the first time she had handled baby wild animals. Ever since she had been a young girl she had devoted her life to caring for animals; two tiny tiger-cuts – this when she and her husband were at Chessington Zoo – and there had been lynxes and young tigers and lions at other zoos. When she and her husband were at London Zoo, she'd looked after "big cats". So Scrap and her brother couldn't have been in better hands.

Of course, it must have been a traumatic experience for the two kits. The shock of Heinz's attack; being wrenched from their mother; these things must have had a devastating effect upon them.

Now it was the day of the official opening of the Highland Wildlife Park, at which Eddie and Joanne Orbell, who was in charge of the children's part of the park, were needed to help with guests invited to lunch and dinner; and they realised there wouldn't be a moment to spare to feed the kits. Colin Burton had the best answer to the problem, he would look after them for the day.

And so Scrap and her brother were added to the kit in the corner, where they huddled together, squeaking and mewing in recognition and greeting each other. While Colin was taking a look at the trio, and realising that he'd now barely have a moment to light a cigarette let alone cook himself a meal, Sanders, all curious to know about the additions to the box in the corner, put in an inquiring nose. It was barely half-an-inch over the edge of the box when it was attacked by a hissing and spitting and flash of tiny claws which sent him back on his haunches with a yip of pained surprise. He looked up at Colin, his expression lugubrious, as if wanting to know what he had done to deserve such a concerted attack upon him, when he, himself, meant nobody any harm.

Colin picked up the box and placed it on the table. He took out Scrap, and noticed at once that she was perky, then he took out the kit who had come with her; it hissed at him, its tiny face contorted. It was bigger than Scrap and the other kit, and he thought it looked in reasonably good shape. It's mother had taught it, instinctively, to fear man and to react to him aggressively. Thousands upon thousands of years, during which man had harassed and slaughtered the wildcat until it had been almost exterminated, had gone into building up that fear. So you couldn't expect it to love you, exactly, could you?

What accounted for this different response in perky little Scrap's case? Simply because, in the first place, Scrap was a female, which usually indicated less aggressive behaviour. And she had been handled by a woman, which would be conducive to a quieter response. Colin was quite entranced by the tiny pansy-like face, the blue eyes glittering appealingly, full of trust. Eddie Orbell had asked him to care for the two kits just for the day, then they could go back to him and his wife. Colin there and then began to wonder if somehow it wasn't going to turn out like that. At any rate, he found himself arguing out loud to Sanders that you could keep three kits together as easily as you could keep one; feeding three hungry mouths was no more difficult than feeding one hungry mouth.

To all of which, it seemed, Sanders barked his entire agreement and, as if to emphasise his full support, attempted again to poke his cold inquisitive nose into the kit's box. Again – hiss, spit, and flash of tiny claws. Again, Sanders' hasty withdrawal and that pained expression, that yip of surprise.

In fact, it so happened that Eddie Orbell quickly spotted that Colin had become devoted to Scrap and her brother at first sight, and although his wife was disappointed at having to give up the kits he decided that they should continue under Colin's care.

Then the first blow struck: shortly after Colin had taken on Scrap and her brother, the first kit died. It happened during the night, and must have been quite sudden. You couldn't tell why, it could have been a kind of delayed shock as a result of the nip Heinz had inflicted on it, from which it hadn't really ever recovered. This often occurs in the case of wild birds, for example, which when they are picked up after being found with a broken wing or broken leg, will last for a couple of days, taking food even. And then suddenly they will weaken and die without any particular reason. It must be assumed that really they have been under shock from the time of the accident, the shock then wearing off and leaving them debilitated, and they just die. It could have been a similar sort of thing with the tiny kit. Colin wasn't altogether surprised at what had happened, though he was disappointed, as was Muriel for him, when she heard the news.

That left Scrap, bright as even, and her brother. Not so bright, though, so far as he was concerned. Where Scrap's brilliant blue eyes were flashing full of life his eyes were still dull.

To prevent any risk of the spread of any unknown virus which may have contributed to its death, the little kit was incinerated. Sanders was awaiting Colin when he returned from this unhappy task, and barked to be taken down to the burn where he swam and paddled about for the next twenty minutes, an amazing performance for a dog half-crippled as he was and with every old bone riddled with arthritis, and with a medical history that read like a veterinary surgeon's casualty department. His leg had been permanently maimed when, aged six, he had been run over by a hit-and-run driver. He had suffered from distemper and enteritis, as well as exposure and near-starvation as the result of an incorrigible predilection for running off and wandering the Cornish roads when he should have been minding cattle. Nevertheless, the last time the veterinary surgeon had come to give him a check-up, the verdict was: "Fit as a fiddle for his age."

Colin always saw to it, and it was becoming a frequent job for Muriel, that Sanders received a thorough towelling after these dips in the burn, to stop his arthritis worsening – unless it was a particularly warm, sunny day when he could be left to drip-dry at his leisure, after which he would retire to his place in the caravan, there to take a long snooze. His most favoured position at these times was to lie on his back, his four paws in the air, his feathery tail giving an occasional wag as he dreamed his happy dreams, as he woofed and snored and twitched the time away.

Muriel often used to take him for a morning walk, and learned to watch out that he didn't charge into a burn or pool where there might be a gin-trap set by some poacher to snare a wildcat, or investigate an old stone dyke, which was also a place where poachers set gins to trap wildcats for their skins, or to have them mounted by taxidermists and sold in shops to tourists – apparently there was an increasing demand for such Victorian-style wild-life trophies.

Sanders had started life as a cattle-dog for a farmer at Colin's home in Penzance, but he would invariably wander off: when he was away for fourteen days at a stretch it decided the farmer that he was more trouble than he was worth. He offered the wanderer to Colin who was working on the farm out of school-hours. Colin was delighted to take on such a strong individual character, and from that moment Sanders hardly ever left his side. Except when there was any water to splash about in; that was ever an irresistible temptation. It was Sanders who in a way, you might say, did a lot to reinforce Colin's love and compassion for animals and helped strengthen his understanding of them. It was during this time when he took on Sanders that his determination to work for and with animals was finally forged, to become the driving-force of his life.

Muriel, of course, came to love him as much as did Colin; Sanders knew it, came to know that he could wrap her round his little finger, figuratively speaking, when it was that exciting moment to be charging off for a swim and paddle in the burn. Grey-whiskered he might be, his eyes glazed with age, but he was still out to prove young blood ran in his veins, and he would demonstrate his aquatic prowess to her until she was quite scared he would drop with exhaustion. Finally, back at the caravan he could flop out in a state of near-collapse while being dried – though he would revive with amazing speed at the mere rustle of his favourite biscuits in the tin.

His attitude towards Scrap, when she would pop her head up at him whenever he went past, was a studied indifference. She would sometimes wave a paw at him, the claws retracted, but Sanders ignored her. He wasn't always going to be allowed to carry on doing so, however. You didn't ignore Scrap when she was around.

Colin had added scraped rabbit to the kits' diet, though the tinned milk, half diluted with warm water, plus the vitamin supplement continued to form the basis of their feeding. Scrap was as enthusiastic as ever about food, though the male kit never did seem to show the same interest.

He remained inert, exhibiting little interest in the business of being fed. Not like Scrap, who behaved as if she was being suckled by her mother – her tiny paws would sometimes reach out as if in that instinctive attempt to knead the mammalian glands to assist the flow of the milk. There were no glands, there was no warm belly into which she could push her forehead to make that special sensory contact. But despite their absence she was receiving some signals of warmth, the warmth of Colin's hands holding the syringe. Signals expressing, in human terms, the affectionate devotion of a human being in whose care she was.

It was in this expression of love that she was learning to place her trust as she sucked from the syringe this sweet, nourishing substitute for her mother's milk.

When she had swallowed all she could take of the mixture from the syringe, and had given her tiny burp or two, she would wriggle as if anxious for comfort. And Colin would hold her against his face, and Scrap would nuzzle against the warmth with her forehead. And press against him with her paws.

It was different with her brother. Perhaps it was significant that he was never given a name, as Scrap had been – she was called Scrap from the start, because she was such a mite, such a scrap, and because she always put up such a scrap, a fight for life. Even during the darkest hour, she fought back, she fought to live. It was this life-force, this electric-like energy that she gave out, that made her so appealing, that entranced you; you couldn't help responding to her. While her nameless brother often acted tetchily, scowled and spat even when he was being fed, she was forever good-tempered. While he hadn't forgotten what his mother had taught him instinctively, that human beings weren't to be trusted and you had reason to fear them, she seemed to be prepared to give human beings the benefit of the doubt.

And so it was that much less easy to feel affectionate towards the male kit. Colin gave him just as much attention, of course, but he wasn't handled so much, held so much, didn't receive the actual bodily warmth that his sister enjoyed. This was to play an important part in bringing them up. Perhaps, even, it was to make the difference between life or death.

Through every night for over a week, the alarm-clock inside Colin's skull would fetch him awake, and he would clamber out of bed and give Scrap and her brother their feed. He was not unused to the routine; his particular concern with the "big cats," babies and adults, at the zoos where he'd worked, leopards, cheetahs and lynx, had been when they were sick. And a wildcat was closer to them than it was to a domestic cat. Its species, was European Wild Cat, and it emerged on the European continent long before man; it hadn't been brought here by traders from Egypt down the Mediterranean as had the domestic cat.

When the hours were tough, nursing tiny creatures like this couple, you'd only be human if you reacted with some cheerful tenderness towards the one who burped pleasurably and purred her grateful thanks for your filling her up, while the other's response, less enthusiastic with often a hiss or baby snarl thrown in, didn't do so much for you. All the same, Colin had plenty of experience nursing baby cubs, and you just had to keep at it, doing the best you could. It was up to you to save the youngster's life in spite of itself.

If you could.

There was a moment when Colin, even his powerful resolve temporarily undermined by lack of a night's sleep – oh, what you'd give to be able to sleep right through – and still with a long day's work to get through in the park, experienced a nagging doubt. Sometimes, when he lay in bed, the wind off the loch gentling the new green leaves of the birches outside their warm, comfortable home, and he would hear a faint mewing from the corner, and he had to ask himself were the odds inexorably lengthening against his being able to rear both kits?

Scrap he felt pretty confident about, she was so full of life, a little dynamo, she was. But the other one? There were going to be a few false dawns ahead, he knew.

As it happened, one or two more than he foresaw.

Chapter Thirteen

Three weeks since the arrival of Scrap and her brother.

After the first week, then, of being fed at two-hourly intervals, then for several days at three-hourly; now it was four-hourly, and Colin by day felt less like a zombie walking around. Though he was still eating too little and smoking too much.

And Scrap's brother was hanging on to life. But he was not putting on weight like his sister; now she was bigger than he was. He still remained apathetic, sometimes it was difficult to persuade him to get his food down. He didn't even respond to Scrap's attempts at playfulness and pats of affection, licking his face with her tiny tongue – sometimes she would lick him quite hard, instinctively, as their mother would have done, in an effort to stimulate his circulation, put life into him.

Scrap continued to thrive not only on her food, but on the way she was being handled. She continued to invite attention by her liveliness, by her readiness to purr away, pushing her forehead against your hands; she loved the warmth of your hands, of being held close to you, cuddled under your chin, against your face. This sympathy and sense of understanding that grew up between her and Colin became increasingly marked – perhaps more so by contrast with the lack of contact he was able to make with the other kit. It was as if Scrap's fur, her tiny tail, were alive with electricity, through which she drew comfort and a feeling of total security. Perhaps she was actually experiencing a human being's affection, even love.

Where Scrap's eyes were sparkling and happy her brother's eyes were hardly ever as bright; usually he seemed to have difficulty in focusing them and they were dull. His fur was nothing like so healthy-looking as hers. It didn't spring back to the touch as hers did. Colin continued to be worried about his lack of steady progress. He still kept them together in their corner of the caravan opposite the one Sanders had taken for his own. Scrap had already house-trained herself, and even her brother, though he preferred to spend most of his time in the straw-lined box, was urged by her, when occasion required, to remove himself to the tray provided.

Realising that her brother wasn't going to be much fun, Scrap soon turned her attention to Sanders, who, since his hostile welcome by the kits when he had first tried to introduce himself, had done his best to ignore her. But she wasn't having any of that. It was his feathery-tipped tail that provided her with most enjoyment.

She would attack it with a sudden pounce and, when he would pull it away with an expression of outraged dignity, she would chase after it, until he withdrew temporarily from the scene. No matter, however, how much she teased him and got under his feet, or however clumsily he tried to ignore her, he never once trod on her, although he seemed to miss her tiny body with his big, floppy paws by a hairsbreadth a dozen times a day.

Colin couldn't work out why, although she and her brother had the same food, for some reason Scrap was getting more nourishment out of hers. Accordingly he went over to nearby Grantown, where the veterinary surgeon who attended the Wildlife Park animals lived, and put him in the picture. The veterinary surgeon approved the quality of the food Colin was using, the half-and-half tinned milk and water, plus the vitamin supplement; and he couldn't understand why both of the kits shouldn't be benefiting equally.

What came up out of their discussion was that perhaps Colin wasn't reading the situation totally accurately in Scrap's case – that is to say, he might be mistaken in believing she was doing all that well. It was possible she might just be making better progress. She was being judged against her brother, who admittedly was coming along on the slow side. If she could improve as a result of a change of diet, this change might also affect her brother to his advantage. At once Colin got the message: the only thing was what other diet would do the trick.

"What about goat's milk?" There was a British Alpine female goat, the veterinary surgeon reminded Colin, in the Wildlife Park. Her name was Heather. She was in milk with no kids to fed; why not milk her and try it on both Scrap and her brother? Colin knew that there wan't anything magical in goat's milk; it's very similar to cow's milk, though richer in fat than in some breeds of cow and less rich than in other breeds. The important factor, however, is that goat's milk fat is broken up into much finer particles than in any cow's milk, and this makes it easier to digest and more readily assimilated into the system.

That same afternoon, accordingly, Colin went to the children's corner of the park where Heather was kept. Joanne Orbell happened to be away, and Colin, though expert enough at milking cows – it was something he had done often for the farmer in Cornwall – thought it might not be quite the same thing dealing with a strange goat. However, it looked as if he would have to try his hand. There are a few small herds of feral goats – domestic goats run wild – to be found in the Highlands, especially in Western Ross, where they are reputed to be of extremely ancient, indigenous stock; black and dark brown, with long straight horns, the points curved, a number of them belonged to the Wildlife Park. With an audience of several children who gathered round, watching with some amusement after a few moments' experimenting Colin soon got the idea; Heather proved perfectly amiable, and a good half-cupful was duly forthcoming.

With this he returned to the caravan and tried it out first on Scrap and then on her brother. Scrap took one good pull at the syringe charged with the new milk and paused only briefly to purr with delight. You could hear her purring from the other side of the caravan, and then in no time at all she had finished the lot. Her brother's reaction was not so marked – certainly he finished off the syringe but his response was less enthusiastic than his sister's. However, it did seem that he enjoyed the new food more than the previous diet and hopes ran high that it might have that extra beneficial effect.

Sanders had looked on with an approving, not to say envious, eye while Colin and Muriel fed the two kits with the new food and whenever he got the chance, which was on several occasions – trust Sanders! – he set Scrap and her brother a good example (well, you could look at it that way) by gobbling up every drop of the goat's milk. Obviously he rated it as great a delicacy as he did ice-cream or vintage rotten bones.

The next few days passed with Scrap blooming on the new diet; her bright eyes even brighter, her fur glistened and was silky to the touch, that characteristic short tail straight up behind her as she showed her perky independence, trotting round the caravan, to play games with Sanders' tail. And now there were crumpled pieces of paper which she was given to pounce on and throw in the air. She always seemed to be purring with delight at the very business of being alive. There was no doubt about it: so far as she was concerned the continued daily allowance of goat's milk had a marvellous effect.

As for her brother, it wasn't quite so encouraging. True, his condition had improved; his eyes were less lack lustre and he did manage to finish his new ration every time he was given it. Occasionally, he would join in with his sister in the games she played, but always he was the first to stop playing to retire to their box, curl up and doze off. Sometimes Scrap would climb in after him, try to rouse him to continue the game. But she never succeeded. While he and Muriel were happy at the enormous difference the goat's milk had made in Scrap's total well-being, Colin still continued to fear for her brother's future.

The kits were now six weeks old. Their tiger-like stripes were beginning to show through increasingly clearly against their tawny bodies. Most characteristic of all was each short, striped tail with its blunt, black "shaving brush" end. They still had their tiny white milk teeth, which would soon, however, be giving place to a complete set of twenty-eight.

The games that Scrap played, and in which her brother sometimes less enthusiastically joined, were all instinctively concerned with hunting. She would sniff the air as if she had caught the scent of her prey – such as a rabbit, or mouse, or vole – and then, slinking with her tiny belly on the floor, she would stealthily approach the spot where the prey was supposed to lie; then at the precise moment she would start to lash her little tail from side to side. Leaping forward in a dramatic pounce she would come down with her claws on the screwed-up ball of paper. Sometimes this would be on the end of a piece of string drawn by Colin to entice her to stalk and pounce. Or, concentrating all her attention on the ball, she would pat it several feet ahead, and in the same moment leap on it with both forepaws, then grip it in her jaws, shaking it from side to side as if to wring the life out of it. Suddenly, she would twist over on her back. all paws extended, holding the crumpled-up bit of paper aloft. She would toss it high in the air and scramble back on her feet again – she was able to accomplish this tricky manœuvre more and more gracefully – without taking her eyes off it for a flicker of a second. As it hit the floor she would pounce on it, send it skittering across the floor and chase after it.

This formed the basis of her daily play, and for variety every now and again she would turn on her brother and pretend to attack him, biting round his neck. He would squeak and mew, and then she would allow him to attack her and she would also squeak and mew defensively, and fight back, her paws lashing the air. And then, of course, there were her forays against Sanders' tail.

But however rough her play, she never showed any signs

of real aggression. There was nothing at all of aggressive behaviour in the way she "attacked" her brother, for example, or the way she pretended to fight Colin's hands, when he played games with her or rolled her on her back to tickle her stomach, which she loved. Always she was purring, and her claws were never fully extended so that she never inflicted any scratches, not even by accident. It was different with her brother, who sometimes would snarl and spit at his sister when he felt the games were a little too rough. Sometimes also he would hiss and bare his tiny teeth at Colin when he wanted to feed him and he didn't feel hungry.

And then that July evening Colin came home and went straight over to Scrap and her brother in their corner of the caravan. Scrap was outside the box, industriously washing herself. Her brother was in the box. A glance was enough, and then Colin bent down and gently picked up the kit. This time there was no attempt at aggression, this time it didn't hiss or scowl, and Colin knew there and then that it was fading. Its life was ebbing away before his eyes. This was the way it could happen with a young wild creature, especially if its start in life had been precarious. Within an hour, even less than that, it could undergo a total change, death could take over without any discernible cause, and its grip fasten remorselessly until life was snuffed out.

Colin watched through the night with the little kit, and as the pinkish pearly grey of the dawn began to creep round the windows Scrap's brother died.

PART THREE

Chapter Fourteen

Colin grieved for Scrap's brother in a way he wouldn't have thought he could; for several days Muriel had to comfort him by reassuring him that he had done all in his power to save the kit. He felt that he ought to have been able to learn what it was that had caused the kit to be sick. But in his heart he was aware there was so much you required to know about wild animals, especially when you kept them in captivity – if you chose that alternative to hitting them over the head and destroying them there and then. That was the choice you had, to destroy them virtually at birth or when they were sick, or attempt to save them, restore them to health.

He was no sentimentalist, the idea of "humanising" animals, attributing human qualities to them, he was totally against; he was very well aware that the little kit's death might have occurred prematurely in any case, while even if it had survived babyhood, it would almost certainly have soon been destroyed by human beings: a wildcat's life expectancy in the wild was only too short. All the same, it took him quite a while to shake off his feeling of frustration and despondency over Scrap's brother's death.

It was Scrap herself who played a part helping him to view what had happened in its right perspective. When he had come back from incinerating her brother he found her, with Muriel watching, walking slowly round the caravan as if searching for him. She was mewing and crying to herself, an odd note that Colin hadn't heard before, a mournful little cry. Muriel knelt and called her, but she didn't seem to hear. You didn't attempt to pick her up to comfort her. You never tried to pick her up; that wasn't the way to deal with Scrap – you always let her come to you, and if she pushed her forehead against your hands and obviously wanted to be held and fondled, then you sat beside her and she climbed aboard, and it was all right.

Scrap went on sniffing round the caravan, while Muriel got some lunch before Colin had to go back to work, and then she came over to Sanders and rubbed against him. He totally ignored her – his attention was expectantly divided between Colin and the biscuit-tin on the table – and even when she purred and pushed herself more closely against him, he continued to ignore her.

Colin bent and fondled her, while Sanders started to bark – to be silenced by a biscuit – and then Scrap, purring noisily, made her way to the box in the corner, sniffed round it and then curled herself up inside it. You could put it this way; she was acting like an animal towards another animal, one of her own kind. So if you wanted to understand her, really to understand her, you just had to get on the same plane with her, identify with her. Once you did that, then her brother's death fell into place; you stopped thinking of him as something human. She was still purring away when Colin finished his coffee and went off to the park.

That night Colin heard Scrap give some little cries as if to herself, she might have been uttering them in her sleep, but that was all – next morning she was scampering about, playing and eating her food as if her dead brother's absence was something she had completely accepted.

Once in a while during the next couple of days she would pause in the middle of a game or while purring to him, and her brilliant eyes would cloud over with a kind of abstraction. Perhaps she was recalling her brother? Wondering what had happened to him? But as the days passed she seemed to forget, she seemed to be caught up so much with her own activities, which centred almost totally around Colin – and she was now getting to know Muriel – and, whether he liked it or not, Sanders.

Of course it was impossible to guess at what thoughts, what emotions were rattling round inside Scrap's skull. She couldn't really have been expected to mourn her brother, she couldn't be expected to know she would never see him again. She couldn't be expected to know that in fact he had died, she couldn't be expected to know there was such a thing as death. She could feel no sorrow, no sense of grief, of loss that her brother, the creature closest to her, had gone forever.

What was it inside her flattened skull, what was it that made Scrap tick? What far away and long ago memories did that small brain hold? Not that it was such a small brain relatively – relatively it was quite large, especially the front part, the learning part, which would be particularly well developed. Normally she would have learned most about life from her mother; after her birth father's only job was to keep as far away as was possible from his family, and his mate would make sure of that, snarling and spitting at the mere scent of him outside the lair. But being torn from home gave Scrap very little chance to learn anything. All that she knew had come entirely through instinct; what she'd "learned" from Colin had really been instinctive – she had known instinctively the games she was playing (which were meant to teach her to hunt), just as, instinctively, she had learned to keep herself clean. The information that reached her brain through her nose, ears and sensory powers, less effectively through her eyes, she must have evaluated instinctively. Picking up the scents daily of Colin and Sanders; and of her home, the caravan, the smells that Colin brought with him from the great unknown beyond the caravan, the scent of the badgers in the park, the wolves and the other birds and animals. And all the scents of cooking, of coffee, of her own food. And hearing the appropriate sounds that accompanied meal-times, and Colin's voice, and sometimes Muriel's, which she knew meant her food was on the way so that she was all ready to gobble it up.

But it wasn't all a matter of instinct. She worked out problems for herself. She experienced enormous comfort and pleasure from the warmth of Colin's body, and since he never did pick her up to fondle her, but always waited for her to come to him, and then sat down beside her, Scrap decided she would get closer to Colin more often – by climbing up where he was sitting, which was often on the long seat running round one side and one end of the caravan. At first Scrap tried jumping up, obtaining a hold with her front claws and hauling herself on to the seat. But it was a bit of an effort, and as Colin watched her – you could almost see her brain clicking over – she soon worked out that by jumping up at the L where the seat formed a right angle at the end of the caravan, she could exert better leverage and pull herself up more easily.

And by her instinct, what did you mean – her memory? A primeval memory, that took her back thousands of years? Could she travel back all those thousands of years which made up her inheritance? The same memory that activated her to store up every item of information her senses picked up? Memory that went back to when her ancestors had prowled the vast forests long before man walked the earth. Then their enemies had been the carnivores of the forests, the bigger cats, the lynx, or the wolf or the bear, before man had exterminated them.

Now, only the wildcat and the fox remained in the glens and hills, on the moors, to be hunted.

It was through this memory, then, that Scrap should have learned to fear man, to react towards him with aggression. But between her and Colin, or Muriel, there was never any feeling of fear, no hint of antipathy towards them. Between them there was only love; or what stood for love in Scrap's reckoning.

Even the brief remembrance of being wrenched from her home – the growling of the dog and the smell of steel – seemed to have had little effect on her so far as her fear of human beings was concerned. Again, you should remember that normally she was largely a nocturnal creature: during daylight you would expect her to naturally behave with a certain amount of tension, you would observe it in her movements, nervousness of manner. But not with her. The day was just as much her friend as the night.

Of course, it might occur to you that, in fact, Scrap owed her life to Colin. But she couldn't know this. There was no reason to imagine that she felt grateful to anyone for being alive. She lived in a world in which human values had no place. She had only her own values, the instinctive values of a wild creature. So however affectionately she purred in Colin's arms, or was to press herself close to him, it wasn't out of gratitude to him for having saved her life.

A life that had hardly begun, anyhow. If she lived, if she did survive, Colin knew he would have one problem to face up to which would require all his considerable determination to solve. It concerned her future. What was to happen to her? A wildcat who didn't run true to form. Not so far at any rate. A wildcat, who should be a snarling bundle of fear for human beings, who should claw and hiss with tigerish ferocity at the sight or smell of them.

But now you're looking into the future.

And you should know better. Even for human beings life was full of hazards, you never knew what lay round the corner. For a baby kit, the risks ahead must be even more desperate.

As when, less than two weeks after her brother died, Scrap went right off her food.

Chapter Fifteen

Colin ate his lunch hurriedly. He wanted to get back to work as soon as he could, since there was plenty to deal with and he would be away first thing tomorrow to Chester on Park business. When he hurried off it hadn't occurred to him that Scrap was anything but her normal self. Only looking back did he remember that it had touched the tips of his antennae, ever-alert, that trouble might be in the air, that possibly she was slightly subdued for her. He would recall noting that she was over near her box most of the time he was having lunch. He would have realised that usually she would have been purring round his legs, pushing against him or trying to clamber up on to his knees, only he happened to have his mind on that work to be done which would leave him free to get away for the Chester business.

When Muriel arrived shortly after he had left, the first thing she noticed was that he had forgotten to take his flask of tea. Then she noticed that Sanders was jittery, he wasn't barking to go down to the burn. She realised it was a thundery afternoon, and she knew how Sanders hated thunderstorms. He would hear them in the distance long before Colin did, and would start trembling with sheer terror. If there was a thunderstorm during the night, he would nuzzle his cold, wet nose into Colin's face and, if it became a really heavy storm, with rain slashing the caravan, the thunder boom-booming overhead and plenty of hissing lightning brightly illuminating the caravan, you would wake up with a start to find this huge, black, hairy shape shaking like a jelly across your head.

In fact, the storm proved to be only a threat; it faded after about an hour. But Sanders still remained on edge, he still didn't want to go down to the burn.

Then Muriel took a good look at Scrap in the corner; when she had first come in she had thought that the kit was enjoying a good snooze – perhaps the thunder affected her, made her sleepy.

But now she realised with a dreadful chill in her stomach that it hadn't been the threat of thunder, after all, that had upset Sanders.

Colin knew as he came back from the park and stood at the door that the kit was fading. He saw Muriel's face, and went over to Scrap, and there wasn't any need for her to say a word to him. This was how it went with a wild animal kept in captivity. You just couldn't foresee that one day, out of a clear blue sky, the creature would be suddenly dying on your hands. This was the way it was, the way it had been with her brother, only the shock was even stronger in her case, since she had always appeared so much brighter than the other kit.

Scrap hadn't had anything to eat or drink since midday – she had now learned to lap up her milk from a saucer, and was eating solid food, bits of raw rabbit – and had taken a liking to Sanders' food, which was often some popular tinned stuff or broken up dog-biscuit. She would nip in from behind while he wasn't looking, and be off with a mouthful without Sanders realising what had happened.

Her eyes looked sunken, her fur was standing up, her nose and mouth were hot. But she wouldn't drink – Muriel had tried her with plain water, since she'd turned away from milk, but she wouldn't touch it. She wouldn't drink now when Colin offered her water.

Was she suffering from an infection passed on to her by her brother? He might, at the last, have been ill with some form of feline distemper, for all you knew it might have been what had caused his death. It is a virus disease; and if Scrap had picked up the most miniscule droplet of the virus off the air, she too could be infected. Incubation could vary from five to six days, but the virus is extremely resistant; it can remain in the atmosphere at room temperature for two or three months.

This is the feline distemper which attacks domestic cats – and it is thought that wildcats can become infected by domestic cats gone feral. But there again wildcats might suffer from a totally different form, or several forms, of the disease. Accurate diagnosis could be difficult enough with domestic cats, which might show similar symptoms for a varied number of diseases: the symptoms for poisoning could appear the same, or coccidiosis symptoms could be mistaken for enteritis (the disease is similar in effect, also, though coccidiosis does not have such devastatingly fatal results). Even enteritis itself comes in various forms, bronchial, for example; and kittens' symptoms can vary again from those of adult cats with the same infection.

And it wasn't as if they would have been able to pinpoint precisely what Scrap's brother had died of. He could have suffered from more than one disease during his continued failing health. It could have started off with coccidiosis, for instance, which could have cleared up of its own accord – it was an infection which often did so – and then the kit, especially vulnerable as a result of the first illness, could have become infected with another disease, such as distemper. Or a disease totally peculiar to wildcats. In their wild state wildcats very probably never fall ill of a virus or other form of sickness, other than that picked up from a domestic animal or some poison deliberately laid by man; if they did, if there was some mysterious natural ailment to which they could succumb, then it would most likely be speedily fatal, and without natural cure, or cure by man. Human beings wouldn't learn of it except by chance.

Scrap, burning up with fever, was rapidly losing weight through becoming dehydrated; that is, suffering from excessive loss of water from her body-tissues. It was essential that Colin got her to drink. He mixed some glucose with warm water, filled the syringe he had used before to feed her and her brother – it had been carefully sterilised – and managed to get her to suck down a few drops of the liquid. He kept on at it, and she sucked down a few more drops. It was the first liquid or food she'd taken for several hours. Glucose not only made the water more palatable and stimulated the tiny taste buds that studded her rough little tongue, mostly at the back of the throat, but is also a vital form of food which, in health, is already circulating in the blood-stream – fever virtually burns it up.

Muriel had got Colin's supper for him before she left, but he felt too sick in the stomach to think of eating. Then he noted the pepper and salt cellars, even as his brain was clicking over fast the way it does when a crisis like this hits, and he reached for some salt. He added a small pinch of it to the glucose and water. Salt is a vitamin essential which fever could burn from the system.

Fortified with cups of coffee, eating his supper in fits and starts, Colin sat up until after midnight, feeding Scrap with the mixture from the syringe. It was a delicate job, drop by drop, and a lot of the fluid dribbling down her jaw. But she did get some of the stuff down her gullet. You could watch her gulp painfully as if to swallow required enormous effort, and your own throat ached in sympathy. Then just before one o'clock, Colin thought he heard Scrap purr. Or was it the sound of the wind under the eaves of the caravan? Or the beat of his own heart drumming, thumping in his ear with anxiety.

He listened again.

It was very faint, but it was a purr. "You were well named 'Scrap'," he muttered. "You're putting up a scrap for your life, all right."

You couldn't help that constriction of the heart, that sudden tightness in the throat; an animal, a dog or a cat, can wrap its tentacles of warmth and love round you, no matter how much you know it's only an animal, that unlike you it doesn't know a thing about personal death, it can't fear it's going to die. And with Scrap, her loss would be even more irreplaceable, difficult to bear – after all, wasn't she unique, wasn't she a wildcat who was different from any other wildcat ever heard of? Who should be on the rampage at even the scent of you, a barbed-wire bundle of spitting, tearing fury. Hadn't she shown an affection for him which no one would believe, hadn't she acted with affection towards Muriel as well, that sort of, well, took your breath away with its innocent trust?

He paced up and down, still holding Scrap, to stretch his muscles, stiffened with tension. He opened the door and sniffed the wind. Sanders was asleep, but with his watch-dog ear cocked and eye opened enough: he grunted and, wide awake in an instant, stood beside him and licked his hand noisily. For once, Colin didn't have to silence his barking. He glanced down at him. Was he quiet out of sympathy – sympathy because he could foretell how it was going to work out?

The air was filled with the sound of the burn below, and Sanders' nose pointed in its direction, his tongue lolling out one side of his jaw. He raised his head for Colin to pat. A tawny owl flew overhead, its ghostly shape slipping noiselessly through the silver birches, their trunks heavy with thick, scaly lichen which shone with an almost phosphorescent gleam in the starlight, and its cry – *Kee-wack . . . Kee-wack . . . Kee-wack . . .* – was answered by its mate somewhere in the direction of the nearby Wildlife Park office-buildings. A pulsing flicker of light began to appear along the top of a bank of cloud behind which hid a crescent of moon.

Colin stared out across to where the loch lay misty and silent in the primevally ancient Vale of Badenoch, the Flooded Land. There came the *Coor-li . . . Coor-li . . .* haunting melancholy from the curlews wading near the loch-shore, and then rising up to plane over the water, their movements accompanied by the soft quack of wild duck in the marshes. With its reedy islands semi-submerged, it was a vast glimmering place of legends and tales of wandering ghosts, sprites and water-kelpies and mysterious hauntings. Colin turned and went back to the caravan, the edges of its windows glimmering welcomingly against the night. He gave Sanders a smack on his behind in appreciation of his sympathetic company, and the dog, with a gentle, growling response, went back to his corner and sprawled out, the tip of his tail wagging affectionately.

Colin held Scrap close, willing her with every ounce of his being to keep on fighting for her life. His jaw tightened, the haggard lines of his face became even grimmer with determination – he wasn't going to lose Scrap, he would save her, come hell or high water; he would, he must pull her back from the abyss. As he reminded himself that he must arrange for someone else on the staff to take on the Chester trip – he didn't intend to leave her until he knew she was past the crisis. He eyed the shelves of books on the other side of the caravan, his veterinary handbooks, his books on wildlife and animals of all kinds, reference-books and encyclopedias, and the lower shelf filled with magazines, and wished with all his heart that if only amongst that mass of information there was some miracle medicine, some magic recipe which would save Scrap's life . . .

His thoughts broke off. He bent his ear closer to her.

Was she still purring – just a little?

He listened intently. You could barely hear her purr, it was that faint.

Chapter Sixteen

Five weeks later.

The phone rang in Jim Taylor's office. He was in the middle of typing a story about a climbing accident the previous evening in the Cairngorms. As he lifted up the receiver he glanced out of the window at the squally morning sky. This was the sort of weather you'd had in the Highlands that summer, sudden bad storms alternating with hot hours of bright sunshine. Taylor knew something about climbing, and as a newspaperman he did his best to try and get over the message that beginner-climbers should take wiser precautions to avoid accidents. As he picked up the receiver he pushed his specs off his eyes onto his forehead. It was Eddie Orbell.

"It's about a wildcat named Scrap. 'Scrap' by name and scrap by nature, she's won a life and death struggle against sickness. I think there's a story for you here."

Taylor knew something about wildcats. From personal experience. You've got to be joking, he thought, as Orbell continued with his account of the wildcat who was so friendly, so loving, towards the people who'd saved her life. He had been grateful to Eddie Orbell the past few months for several stories which had come out of the Highland Wildlife Park. One of the best in the business, he was, highly experienced, knowledgeable and always extremely helpful over stories about animals and wild life.

His enthusiasm for this baby wildcat, named Scrap, caught Taylor's imagination. It had been brought up by Colin Burton, the Wildlife Park's head warden – he'd recently married Muriel Bishop, who'd been helping him with Scrap, as a result of which they'd had no time for a honeymoon. Jim Taylor had met her and Colin at the late-night cinema in Aviemore; though her home was near London, her mother was Scottish. He would be over, he said to Eddie Orbell.

After he had hung up it occurred to him that this would make a good story for Scottish BBC Radio – not only did he run this news agency which covered the entire Highlands area, but he was becoming known as a reporter for radio and television. It seemed just the story that would interest the BBC. But he was feeling more than a bit dubious about this wildcat.

He had a vivid picture in his mind of that time only a few months ago when he had been out for a climb up in the Monadhliath Mountains, which were divided from the Cairngorms by the fast-flowing River Spey. He was making his way down when a sudden blizzard hit the area cutting out visibility almost entirely. He knew where he was, however, and had enough sense as a result of many similar experiences to find shelter, wait until the blizzard stopped. He had plenty of daylight left in which to wait for it to subside so that he could get home.

He clambered down between an outcrop of rock and a quiet spot and let the blizzard howl to hell over his head. He had just untied his climbing-axe from his wristband, and slid out his feet to undo his pack for some pork-pie and a thermos of coffee, when there was a sudden whoosh – spit and scream – then a wildcat zoomed out from under the rock-ledge just by his boots and flew at him. He ducked and dodged and it shot off into the blizzard, spitting defiance and ferocity as it went.

The wildcat must have been there when he first got down into the rocks and must have lain quiet until he had started to undo his pack. This must have given it the impression that it was about to be attacked. Taylor sat both shocked and interested at this near-miss of being attacked by a wildcat. When the blizzard eased he moved down the mountainside in the direction that he figured it had taken but saw no sign of it, and the snow had covered its tracks.

Jim Taylor was quite certain the idea of taming a wildcat was out of the question. It had been tried several times but no one had ever succeeded. But Eddie Orbell had promised him that this one was different from any other that had ever been known, and he'd promised that he would come to no harm. All the same he searched a drawer for a thick pair of gloves, and found the pair he'd worn that time on the mountain.

As he left Aviemore for Kincraig early that afternoon, more squally weather broke. There had been an intermission of warm sunshine but that was now breaking up and the wind rattled his car-windows, and he soon had to have the windscreen wipers going. Jim Taylor knew this part of the Highlands as well as any gamekeeper, poacher or shepherd. He knew the rugged terrain of glens and lochs, of mountain burns, with their plump trout and silver salmon, he knew the bens and hills where the deer roamed, where fox and otter lived; and the mountains where golden eagle, buzzard, peregrine falcon, merlin and hovering kestrel could be seen and where, around the three thousand foot contour, ptarmigan and mountain hare changed their colours according to the season and the surroundings. He knew and loved not only the absolute splendour of the wildness and majesty of the mountains, of the beauty of the forests and the glens, but he knew Aviemore, become famous as a holiday centre with its huge modern hotels, its ski-slopes and centres for climbing, riding and pony-trekking. It was an all-the-year-round season: always plenty of stories to keep him on the go. It was a life of marvellous contrast, which he reported.

He turned off by the side of the glen and up the road towards the Wildlife Park offices, heading for Colin Burton's

BIRDS

) plays with
.'s jacket at home
e Burtons'
an

with Colin
on

caravan, his mind filled again with his previous encounter with a wildcat. It had been an experience which had scared him off wildcats, which was why he was still filled with a certain amount of apprehension as he stopped his car by the side of the glen, where the birches waved their branches in the squall. He heard Sanders barking in the caravan. He had met him, too, this old black, lame dog who was full of enthusiasm for life – especially splashing and paddling in the burn down below.

He grabbed his recording equipment and camera and made his way to the caravan. He still had in his mind the picture of that black striped, tawny body which showed up even in that blizzard and the wildcat's glistening teeth, and flashing, slitted eyes, ears flat against the top of its skull, its claws reaching out towards him so that he had to move fast to get out of their way.

As he hesitated, the caravan-door opened, and Colin Burton greeted him. Muriel – now Mrs Burton – was getting him a cup of coffee. But he was staring disbelievingly at a sight he had never expected to see as long as he lived – this wildcat playing games with Sanders' feathery tail. The big black dog obviously wasn't minding; in fact as you watched, he opened his jaws in a great yawn, and this wildcat stopped its game for a moment to stare at the glistening teeth, as if realising that it could be snapped in two with one great bite. Then it went on with the game without a flicker of fear.

Chapter Seventeen

Scrap had caught the sound of Jim Taylor's car several moments before Sanders had. Her hearing was much more acute than his, because he was getting on a bit in years. And she had picked up this strange new scent as Taylor approached the caravan. And as he came in and Sanders barked she turned and stared at him with intense curiosity.

She couldn't know that for several weeks she had been coming through the valley of death. Since she couldn't know what death was, it meant nothing to her and so she never had any fear of the future. During her illness she had experienced no anxiety that she would die. She had lived from moment to moment, feeling very ill, experiencing pain, the burning sensation throughout her entire body; every fibre of her being had burned, every nerve had been on fire, every muscle. And her nose and mouth had burned so that she had endured that raging thirst until it was quenched by the sweetness of the liquid in her mouth.

She had felt the sickness in her little stomach, although she had not known how long it had lasted; then she had felt that sickness, too, begin to recede. She had felt the cool of Colin's touch, its tenderness reminding her of her mother, whose memory was fading. Soon that memory would vanish altogether: if she had ever met her mother again she would not have known her.

Her illness had meant a time of sleeping and waking, sometimes half awake only, sometimes dozing. But she had felt the strength gradually returning to her limbs. The nights and days had passed in a continuous blur of memories and scents, smells of cooking, the scent of Colin as he returned to the caravan from work every day; and the scent of Sanders. The sound of his barking and grumbling when Colin rebuked him, or when Muriel spoke to him sharply. And she caught the change in Colin's or Muriel's voice after they had spoken sharply to Sanders – that sharpness was replaced by the tone of affection, the same sort of tone which they used when they spoke to her.

Vaguely, dimly, she had heard other sounds. The sounds outside the caravan of storms, of rain beating on the roof, of the wind whistling round the eaves, of the birds singing when the sun and warmth of the spring day emerged from behind the black clouds and squally weather. In the long nights the caravan became more silent, Sanders asleep, growling every now and again in his dreams, and she had taken comfort from his wuffling and sleepy grumbling. And there had been memories of those games with his feathery tail. And the only sound might be the hoot of a tawny owl – except always she could hear the murmur of the burn in the glen and she could hear the scrape of the branches of the birch-trees. She collected these sounds and scents in her memory until they became a part of her.

She did not see so much at the beginning of her illness. Her eyes had changed from the blue of a kit to that of a growing wildcat, now they were a golden, brassy colour with the characteristic black pupil slitted during the daytime. And it was during the daytime that she found difficulty in seeing, especially when the sun shone brightly into the caravan. She could make out the dim shadowy figure of Colin or Muriel and the dark figure of Sanders.

When the curtains were drawn against the night and there was only the glow of lamplight then Scrap's eyes were better able to focus, the vertical slit pupil expanded, so that in the shadow of the box where she spent her time when she wasn't being handled and fed by Colin, her eyes glowed up at you brilliantly like flashing jewels. At the beginning of her illness her third eyelid, or the nictitating membrane, had shown itself. This was due to her illness and its use was to protect the eye against anything entering it. But as she got better this third eyelid disappeared.

Scrap's eyes were in fact a kind of compromise, you might say, adapted so that she could hunt both by day and night. A pupil shaped like hers could be opened more widely and closed more narrowly than the round pupil of, say a human eye. And a vertical slit was of more use than a horizontal one, because in bright light the amount of light entering her eye was still further reduced when she partly closed her eyelids, thus reducing the functional part of the pupil to a mere pinpoint. What made her eyes glow at night was that beneath the sensitive nerve-cells inside her eye was a reflecting layer, so that in dim light each light ray had two chances of affecting the nerve-cells which in their turn controlled vision, one ray of light on the way into the eye and the other ray of light on the way out again.

Had the fever during her illness resulted in Scrap suffering moments of delirium intermixed with lucid moments? Had she played again those games with Colin; had she dreamed of hunting, stalking her prey hidden there under the table in the shadows? Or had she remembered none of these things but simple remained suspended in time almost, instinctively conserving her energy, not wasting it even on dreaming of past enjoyments?

Had she known on that second night of fever, as Colin had known, that she had reached the crisis?

It had been about two a.m. It was obviously still touch and go, Colin had kept up the glucose and warm water with the addition of a pinch of salt. But had Scrap's fight for life proved too much for her strength? She had taken the nourishment, such as it was, for over forty-eight hours, being fed non-stop by

Colin. The fever had abated, it seemed certain, but during the time it had raged had Scrap lost the strength with which to continue to live? There were times, Colin admitted to Jim Taylor, when he thought she wouldn't make it. He wouldn't admit it to himself; then, two a.m. by his watch, he caught a sign – "A tiny movement of her throat." Could it really be, he wondered as the movement ceased, or was he just imagining it? Then he saw the tiny throat expand again, "just for a fraction of a second".

His heart surged with hope. Scrap took a little more of the glucose and warm water. He saw her throat move again. Slowly, agonisingly slowly, the throat movement continued. The glucose was now going down and Scrap was getting some strength back at last.

Scrap experiencing the return of strength and feeling to her legs, the wish to stagger to her feet where before there had been no wish and even less strength; tentatively licking the tip of her nose, instinctively moistening it to help pick up the scents of Colin and Sanders, and of Muriel; the scents of the caravan. Then when the door opened – Colin or Sanders going out, or coming in, or Muriel arriving – picking up the scents and sounds of the outside world. That world into which she had never yet properly ventured. The world of trees and sky and birds singing, the burn down below in the glen. Scrap becoming stronger every day, experiencing to the full the joy of being alive.

Now, she was hearing this new voice; now, her scenting power, which was back to normal, caught this strange new scent of some human being who had just come in, with Sanders barking. And she knew by the tone in the voices of Colin and Muriel that this human being was a friend of theirs, and was therefore a friend of hers. She leapt on to the seat running along the side of the caravan, at the other end of which Taylor had placed his radio recording-gear and his camera. These were new smells for her.

Colin and Muriel watched her as, purring loudly, she greeted this stranger – you couldn't credit it, her trust and friendliness – and Jim Taylor stared at her in total disbelief. He had taken off his gloves, and now she licked his hand with her rough, pink little tongue.

Chapter Eighteen

Jim Taylor had allowed his panatella to go out.

It was a sensation he was never to forget, this warm, rough, little tongue licking his fingers as if to say Scrap hoped to be his friend. He reached for his camera and flash. He was nervous that the flashlight might scare her off, might frighten her. Just to try her out he flicked off the first shots away from her in case she would panic. But not a bit of it, she took to it as if she was quite used to being photographed by flashlight. He went ahead shooting as many pictures as he could get of her, first of all with Muriel, then with Colin, then Colin and Muriel together. Then, of course, with Sanders, who obliged with a certain perfunctory nonchalance as if he wondered what all the fuss was about, every now and again letting loose with a burst of barking.

As Jim Taylor got his pictures Colin went on to describe how Scrap had come to them, how her brother had died; and how he wished he had managed to save him. Then of Scrap herself falling ill.

While Colin was talking about Scrap's brother's death, it seemed to Taylor that he was still disturbed about that loss, that he wished fervently that he had been able to save him, holding that little creature and knowing that its life was slipping away through his fingers even as he held it, that tiny spark of life flickering out. You might say it was as if it was a blow to his professional pride. He hadn't really accepted the failure to save that one life while he had saved the other. He still had it at the back of his mind that her brother should have survived as well.

While Scrap settled herself comfortably on his cap and gloves, purring away and looking up at him gratefully, so that he could hardly take his eyes off her, Jim Taylor spoke about Gilbert Fisher, when he had been director of the Edinburgh Zoo, and his experiences with two wildcat kits he had taken care of. Though the pair of them had grown up to be reasonably friendly with him, they'd sometimes scowled and spat even as tiny kits and struck out with their paws at him, while to strangers they always exhibited typical wildcat fear and aggression. Fisher had kept them for several weeks in his own home. Then he'd realised they were growing increasingly aggressive, especially towards anyone else. He had decided therefore the best thing to do was put them into his zoo, which was what he had done. They quickly became like any wildcat in the wilds, ferocious-looking, hissing and spitting fury at the sight of human beings.

And there was the dramatic night a couple of years back, Taylor was reminded of, when a gamekeeper over near Loch Ness at a place called Drumnadrochit, beyond Inverness, had shot a wildcat, to discover it was a female and she was pregnant. Curious to know if her young were alive he had opened her up with his penknife there and then. There were four tiny kits – alive. The gamekeeper remembered Mrs Joyce MacDonald, a doctor's widow, who lived in an old cottage on the hillside at Drumnadrochit, who had saved the lives of several wild creatures, fox-cubs and birds, and he guessed that she would very much like to try and save these kits. He drove through a terrific storm raging at the time, and managed to get up the hillside to Mrs MacDonald, but by now one of the kits was dead. The other three were just alive. Mrs MacDonald had taken them on. Although two died a few days later, one survived, which she named Angus.

Mrs MacDonald reared Angus for several months, and although he had become friendly with her, towards anyone else he would scowl and spit. At any rate, when he was aged only nine months, he had died from cat enteritis.

So it was impossible to know, Jim Taylor said, how long Angus would have remained friendly, even towards Mrs MacDonald.

He knew how gamekeepers and shepherds swore you couldn't tame a wildcat; naturalists were equally convinced – it would always be ferocious towards man. Yet here was Scrap, as friendly as could be to him, acting as if she knew exactly what was required of her. Going through her repertoire of games especially for his benefit, you could say, with Colin and Muriel, playing hide-and-seek and chasing after the ball of paper, and pouncing on Sanders' tail, which he would whisk away while she still chased after it. Sanders even obliged by putting on his most bored expression while he allowed Scrap to go through her performance of stealing his food. By now he had realised what was going on whenever he turned his back on her, but this afternoon he pretended to be taken completely by surprise. Inevitably he ended up with a good fit of barking.

What was so appealing about Scrap was this expression she gave of total friendliness and affection towards Colin and Muriel. It was extraordinary to watch her almost talk to them with purrs, licking their fingers, and always putting out an outstretched paw, not as Taylor remembered the paws of the wildcat that had gone for him, flashing claws extended, but with her paw-pad patting their hands or Colin's face. He was watching Scrap and Sanders mixing it in a mock battle, then she would pretend to leap to safety to Colin's lap. And then jumping down again and confusing poor, patient old Sanders with another mock attack on him.

Jim Taylor knew he'd got a great animal story. He knew that he'd got a really strong radio broadcast filled with human interest and drama.

Scrap inspected his microphone and recording-gear as he prepared to interview Colin and Muriel. This was a new smell for her, this metallic scent, and her nose twitched with deepest interest as she sniffed away. Then, and he could scarcely credit his ears, she edged herself in between Colin and Muriel as they were talking into the microphone and began purring. It really made a wonderful climax to the interview, and he said into the microphone: "Now you can hear Scrap herself expressing her gratitude to Colin and Muriel," and Scrap duly purred even more loudly.

When the interview ended Taylor's hand was trembling with excitement, and when he took another panatella, it was a moment before he could light up.

Old Sanders had slumped in his corner, quite exhausted by the carryings-on, while Scrap came over to Jim Taylor as he was collecting his gear, just to supervise that everything was in order, patting each item, and purring good-bye to him.

Chapter Nineteen

Scrap was now aged nearly four months.

In her natural state she would have been weaned by now, she would have left her mother's side to go off with her brothers, playing games, learning to hunt for food. Learning the lore of the wild.

In a further six months she would be fully grown. At a year old she would be ready to mate.

For the past couple of weeks, Colin had been talking to Muriel about putting Scrap back into the wild. It was in the wild where she really belonged. His ideas were quite clear, it had always been his intention to return Scrap to her natural environment. Muriel, of course, knew this. She had known very well that the time would come when Scrap would have to leave. Leave her and Colin, leave her home in the caravan, to spend her days in the Wildlife Park. In the wildcat enclosure, already prepared, which was an extensive area, with rocks and trees and a pool – a fine example of the sort of environment suitable to wildcats.

Muriel may have entertained the vaguest idea at the very back of her mind that Scrap could have stayed with them, permanently, but, of course, it wouldn't have been right. She went by what Colin said. He said that it was totally wrong to attempt to "tame" a wild creature, to domesticate it. The fate of a wild animal kept as pet was always the same. Forced to live a life for which it was never intended, ending up sick, to be destroyed; or to ending up in a zoo. Muriel knew Colin's plan was to put Scrap as close to the wild as makes no matter, while yet giving her a security that she could never have enjoyed if let loose in the forest.

Sooner or later, there would be more wildcats to keep her company. They were not easy to come by – a wildcat was shot on sight – but there were gamekeepers such as Hamish Munro who would keep a look-out for a wildcat to bring to the Wildlife Park. They were most likely to be young kits. So Scrap could look forward to companionship. She wouldn't be alone in the wild.

However, before he put her in the park, Colin had decided that she should first live in a small enclosure near the caravan, specially made for her so that she could prepare herself for her new home, while at the same time she would be near him, and Muriel and Sanders, so that they could keep an eye on her. She would keep in contact with them, while at the same time she would become acclimatised to the sort of environment she would live in later.

So far, she had spent almost her entire life indoors, warm and cosy. Quite different from what she would face when she went wild.

Colin built a special wire-netting enclosure for her. It was completely secure, and she had wooden sleeping-quarters at one end of the enclosure. Her first lair, as you might call it. Warm and rainproof, with always plenty of hay to curl up in.

The enclosure was built only a few yards away from the caravan, to which Scrap could be brought as often as she liked. She would be weaned away from her home as gently as possible. This was Colin's idea. He had a pretty good notion that Muriel would want Scrap with her often. But at any rate she would be spending the night-time out of doors, which was the time she would come to know best; she would have her own home. She would be learning what it would be like to spend the rest of her life out-of-doors. The sort of life she would have to become accustomed to.

One morning, then, accompanied by Muriel – and, of course, Sanders, barking as if he was aware it was some special occasion – Colin took Scrap out of the caravan for the first time in her life. She snuggled up in his arms as he carried her to the enclosure, all brand-new wood and shining wire-netting, and her wooden sleeping-quarters with fresh hay, all ready for her. Colin opened the enclosure door and popped her inside. The door was bolted and padlocked after her. Colin had to go off to the park, and he left Muriel to stay behind with Sanders, talking to Scrap for half-an-hour or so. Then she, too, returned to the caravan with Sanders. Scrap was left walking round her new territory, tail up, and purring with obvious satisfaction about her surroundings.

She took to her new home like a duck to water. She spent most of every day in her enclosure, some of the time sleeping hidden away in her lair, but early each evening she would be brought to the caravan. If it was an especially fine evening, Colin and Muriel would let her outside the enclosure, where she would play games in the grass, sometimes disappearing with Sanders under the caravan. Playing games was, of course, important to her, as it is with most young animals. She was building up her muscles, shaping her reflexes; and so Colin saw to it that she had plenty of time for playing. With him and Muriel, sometimes with Sanders; and sometimes she would race about on her own.

Colin reckoned she would require to stay in her new home for about a month or two, and then she could be taken up to the wildcat enclosure permanently.

Muriel foresaw that, when this happened, there would be less chances of seeing Scrap, and so she would have liked her to spend even more time in the caravan, but she tried not to bring her in too much – she realised that Colin's objective was that Scrap should learn as quickly as possible to live out-of-doors, against the time when she would virtually be in the wild. But she was such wonderful company, and Muriel missed her. Although she didn't make anything of it to Colin, she wasn't at all looking forward to the day when Scrap went away.

Meanwhile she enjoyed every moment with Scrap; there were several occasions when she was bringing her to the caravan when she would deliberately "escape" from her arms, to dash underneath the caravan for a hide-and-seek game with her, and with Sanders. Or if the caravan-door happened to be left open for a moment, she would get out, and the same fun-and-games would follow. Muriel would have a fine time trying to catch her, before Scrap finally allowed herself to be captured and it was back to the caravan.

Scrap had a good view from her new home of the birches by the glen that led to the burn whose sound she had heard from the caravan, she could sniff their scent and that of the resinous pines; the dozen varied smells of the earth itself. There was the open sky above, summer bright sometimes, or sometimes clouded with storm – and she had shelter from the rain; there was the flight of the blackbird and its warning screech of its song. The clatter of wild-duck from the loch, and the *coor-li – coorli* of the curlew – all these sounds she had heard from indoors, but now she was hearing them in true perspective. She could relate the warning thump of a rabbit's hindlegs to the sight of it dashing past the enclosure, eyes abulge.

By night she could find herself in a new world. Of starlight, of moonlight, a world of light and deepest shadow, when other nocturnal creatures came to life, the fox, barking to his mate in the distance; the tawny owl, a silent phantom overhead. All these things Scrap quickly learned to understand, she registered them instinctively, the sounds, the scents, the shapes she observed.

And there was still that other world she already knew. The human beings' world.

One afternoon Muriel returned from shopping in Kingussie and as she approached the caravan she heard the sound of children's voices. It wasn't that they sounded so loud or excited, but that they sounded full of wonder, as if they were tremendously excited, but were afraid to appear so. Their voices came from the direction of Scrap's enclosure. Muriel quickly parked her shopping in the caravan without disturbing Sanders from his afternoon snooze, and hurried over.

There she found a crowd of half-a-dozen children gathered round, talking excitedly to Scrap through the wire-netting.

For a heart-stopping moment Muriel was scared that Scrap might have taken fright and exhibited fear and aggression towards the children – they were, she saw at once, poking their fingers between the wire netting. But when she reached the enclosure she saw that Scrap wasn't in the least bit afraid but was purring away in her usual manner and thoroughly enjoying the company of the children; she was licking their fingers through the wire-netting to their shrill delight. It was at this moment that Colin appeared on the scene.

It appeared that the youngsters had heard Jim Taylor's radio broadcast with him and Muriel, during which Scrap had contributed her performance. They had also read about her in the newspapers. The children had remembered this when they came to the Wildlife Park to see the other animals. They wanted to see this wildcat living in the caravan, then when they had found no one there, they had discovered Scrap in her new home and quickly made friends with her.

They wanted Colin to take Scrap out of the enclosure and let them hold her. But he explained that perhaps it might frighten her. He couldn't help thinking that Scrap appeared as if she would have been only too glad to be held by the children, but on the other hand he felt that perhaps it might be too much for her, she could become frightened and aggressive and attack the children.

Though anything less aggressive, you couldn't imagine, as you watched Scrap purring and pushing against the wire-netting in an effort to get as close as she could to the children's warmth. It was because of her extraordinary friendliness, even affection, towards children, that there was obviously no risk of her suddenly turning on them; thus she was to be moved into the children's corner of the park. She would become an increasingly popular attraction, making friends with hundreds of youngsters, who were to delight in petting her and listening to her purring her enjoyment of their company.

Muriel awoke to the sound of a storm lashing the caravan. Colin came to at the same time, yawning his displeasure at the noise that had awakened him. Together they lay there listening to the crash of the wind against their home, then Colin answered Muriel's unspoken question. "I'll go and see if she's all right," he said. He pulled on some clothes and went out to, inevitably, the noise of Sanders' barking. He accompanied Colin to the enclosure, limping arthritically as if to point out that this weather didn't do much for him, either.

Scrap had retired to her warm, dry shelter. Colin called to her and, seeing him there, she poked her head out, her eyes glowing in the darkness, to utter a kind of grumbling miaouw. She didn't enjoy rain.

It was certainly lashing Colin as he stood there. He said goodnight to her and while Sanders went back with him to the caravan, returned to report to Muriel that all was well. They finally fell asleep again, but Colin was woken again as the dawn broke with the storm still continuing by the crash of something outside the caravan. Immediately afterwards Sanders started barking and Colin hurried to the door to look out and see if anything was wrong. From the door he could see Scrap's enclosure. To his dismay he saw that the wire-netting structure had been blown over on one side. The wind, gusting hard, had somehow caught it and lifted it a couple of yards.

From where he stood there was no sign of Scrap.

Muriel was awake by now, asking what the trouble was, and while Colin put on some clothes, he told her. She herself dressed quickly, and they both went out into the storm. Sanders rushed ahead barking loudly but he seemed to have no idea where Scrap had got to. At once Colin and Muriel began searching. They went down into the glen where the burn was, and for once Sanders refrained from splashing about or enjoying a swim.

The trouble was they didn't know exactly when, in fact, Scrap had found herself free. The crash which had awoken Colin could have been caused by the wind with sudden extra force smashing against the caravan or slashing its way through the trees. So that the enclosure might have been blown over some time earlier on, giving Scrap plenty of time to escape and get some distance away.

With harassed faces streaming with rain, and even Sanders quiet, not once barking in his exuberant way.

The most worrying thing to Colin and Muriel was that Scrap might have bolted in sheer blind panic – if she hadn't been injured when the enclosure turned over, and had crawled off to hide and lick her wounds. They might never find her again. Colin and Muriel searched for the next two hours. Colin hurried down as far as the old road, perhaps Scrap had even got as far as that. He called "Scrap" all the way. He searched along the road, dreading seeing her lying there, struck by a passing car. There was no sign of her, at least she hadn't been killed that way. Clouds moved slowly over the loch, trailing grey streamers of vapour across a pink sky, shot with blue, reflected in the rain-slashed surface. Across the loch, black against the dawn, Inshriach Forest stretched for miles into the rain-mist, and beyond curved the hills of Rothiemurchus, and beyond them, the Cairngorms, like great rollers breaking on the far storm-battered horizon. He went under the railway-line to Inverness, the walls of the short tunnel streamed with water.

The rain rattled down on the corrugated iron roof, red and rusted, of the boathouse, lying so low in the marshy shore that it appeared to have become embedded in the reeds. Colin called and searched carefully – Scrap might have found a hiding-place hereabouts. He stared out where the narrow channel from the boathouse ran past some thickly-clumped reeds, from which with a flurry of wings, a pair of wild-duck took off. But it was he who had disturbed them, not Scrap. He searched all round a rotting fishing-dinghy, ribs black and slimy, buried in the mossy coarse grass and some small rocks. He knew that Scrap could play hide-and-seek up to the last minute before condescending to give herself up.

With Sanders close beside him, he turned back. The railway-line telegraph-posts leaned unsteadily against the squalls as he went under the bridge again and recrossed the road.

Meanwhile . . . Muriel had given her urgent attention to the area of the park offices and outbuildings. Knowing Scrap's dislike of rain and fearing, too, that she had panicked and rushed off in the first direction that took her head, Muriel hoped she would have sought shelter. By way of an open window, or a door left open. She looked into every doorway, tested every door in case it had been open, then slammed shut. She was drenched, her hair sticking to her face, as she battled her way round the buildings. Not a mew, not the faintest sign of Scrap.

She reached the last building. "Scrap . . . Scrap . . ." she was shouting, and then buffeted by an extra vicious squall, she had passed the open doors of a barn, where hay and foodstuffs were stored. She reeled under the impact of the lash of rain, and was about to push on when she heard Colin shouting.

He had been passing the stables on his way back, Sanders tagging alongside, when suddenly in a lull in the roar of the gale, he heard an unmistakable mewing. It must be Scrap. It could only be her. He cut short Sanders as he started to bark, and calling out to Scrap, moved towards the stable-door. As he did so, out she came, tail erect, purring unconcernedly, bright as a button. Sanders reacted with friendly woofs, and she brushed past him to arch her back against Colin's legs.

Shouting to Muriel that he had found Scrap, Colin looked into the stable, to see where she had been hiding. And there was a baby red deer, still asleep in a "nest" of hay – she was an orphan, found on the hill, and brought in to safety, to be hand-reared. Scrap trotted over to the wee dappled creature and licked its face with her little pink tongue, as she curled herself beside it. This was where she had spent the night, warm and cosy against the storm outside.

Muriel stood next to Colin, staring down, while Scrap yawned, stirred herself, stood up and came forward – it was an incredible sight – you just couldn't credit it, ever. A wildcat who had curled up comfortably and sleepily with a baby red deer. In the natural way they were mortal enemies.

Muriel gave a cry of thankfulness and was about to reach down and pick Scrap up to hold her in her arms. Then she stopped, remembering. You always let Scrap come to you. And even in that moment of excitement, it occurred to her – and it was something that she was later to have reason to recall – that to attempt to make a movement like that when she was half-asleep and not expecting it might cause Scrap to react with fright and consequently turn on her.

Muriel was on her knees beside her, talking to her quietly. Scrap began purring that deep, throaty purr that most nearly resembled a steam-locomotive than anything else. She yawned again hugely, stretched herself luxuriously, and came to Muriel, that tail up in the air. Muriel held her to her face. Sanders began barking, Colin said something about her being soaked to the skin, but Muriel couldn't answer – she was laughing with happiness – Scrap was busily licking the rain from her face. And the tears of love.

Chapter Twenty

Now, you're realising what a beautiful creature Scrap was becoming. Every now and again she would stop the game she would be playing, tu walk round the caravan, tail erect, to signal her pleasure at being with her three loved friends. Her coat was a wonderful tawny colour with typical black "tiger" stripes (as she grew basic tawniness would increase); and her face a characteristic wildcat mask, beautiful tawny-gold eyes. And that marvellous black shaving-brush end to her stubby tail. As you watched her she would crouch, extending her fore-paws, and stretch herself. She would push her paws out so that you could see the retractable claws.

Colin would watch her turn her brilliant golden eyes on Muriel full of warm affection, and he couldn't help reflecting how different they would look slitted in wildcat aggressiveness.

Apart from helping in the general camouflage effect her stripes were meant to be used to terrify other animals she might encounter. You can imagine a wildcat, for example, meeting a rabbit in the darkness of the forest, teeth bared and the stripes round the eyes drawn back distorting them fearsomely. The rabbit would be literally petrified with fright and before it knew what had happened the wildcat would have pounced.

Even watching her strutting round the room and purring away, you could have in your mind a very different image from the one which Scrap was conveying right now. Colin had warned Muriel of what had happened before with the two or three wildcats so far brought up in captivity. The kit would form some kind of attachment with the human being who looked after it, but sooner or later it had reverted back to nature. It had turned upon its owner without any warning, to scratch and tear with its formidable claws, and it had been found impossible to keep it in captivity any more.

Would this be what would happen in Scrap's case? Would she inevitably turn wild? Despite all the love which she seemed to hold now for Muriel and himself wouldn't the time come when that love and trustful affection must turn to aggressive fear?

When you are thinking about Scrap, wondering about her, you are dealing with a wildcat who was living a life unnatural to her. Everything she had learned, and was learning, was mostly from Colin. Plus what she already knew instinctively. And what she already knew instinctively was very important to her. The only thing was there must have been some break, some twist, in her instinctive knowledge, because she should fear man, she should act aggressively towards man. She should snarl, hiss, spit, flatten her ears, crouch ready to spring to the attack at the approach of any human being.

But she did nothing of the sort, she was acting completely out of character. Fear of man, which was part of her evolution, which must have been ingrained in her instinctively ever since man walked the earth and hunted a million years ago, this instinct had broken down somewhere along the line.

So now you are dealing with a freak wildcat.

Scrap was unique.

There had never been another wildcat like her. What had been responsible for this breakdown in her instinctive behaviour? It must have begun with feeding her. Especially when she was ill. She had always been handled much more than her brother had been handled. During her illness the closeness between her and Colin had been intensified. He must have conveyed to her that she was secure, that she had no reason to fear him. That she could trust him totally. By the touch of his hands against her body, for it was as if she drew messages of love and trust through his hands.

She was always being stroked by Colin and Muriel; she invited this kind of attention by her attractive liveliness, her eagerness for attention, she wanted to be touched, to be fondled. She responded with mewing and rubbing her face and head against Colin's and Muriel's hands. She loved the warmth of hands, she loved Muriel's face and neck and Muriel's and Colin's body warmth. She thrived on touch, on the warmth of hands and body. Her sensory powers were developed beyond human imagination; her fur and tail must have been alive with electricity; her whiskers' ends were the same as a man's fingertips. Only in Scrap's case their sensory ability was a thousand times more highly developed, more subtle, and moreover worked much more quickly than the fingertips of a human being.

And then, early one mid-August evening.

It was very still; you could hear the murmur of the burn down in the glen distinctly – Sanders' ears as he snoozed in the corner twitched as if he was hearing the sound in his dreams, and you could imagine him any moment leaping up barking to be allowed out for a splash and a swim. Through the window, a solitary star in the hazy blue of the sky.

Muriel was expecting her husband home at any minute. Scrap was with her; they were playing the paper-ball game, Muriel throwing it into the air for her to catch – sometimes Scrap would alternate catching it by allowing it to bounce, which it did slightly, and then pouncing on it. Sometimes – and you could believe she did it "accidentally-on-purpose" – she would flick the ball on to Sanders snoozing in his favourite position on the seat. And she would leap up on to him, not always, you might think, with due regard for what part of his anatomy she might land on, and he would give an agonised grunt, but she was off, out of the way of a protesting paw. This time, she let the crumpled ball of paper hit the floor, but for a fraction of a second she was put off her stroke. For at that precise moment, Sanders gave an enormous yawn, at the same time allowing his tail to flip lazily over the seat edge, and interrupted Scrap's concentration on the ball. She missed it as it bounced, instead only half catching it with a paw, to send it skittering into a corner.

Watching her as she decided to chase after it, instead of going for Sanders' tail, which he anyway, only too well aware of Scrap's fondness for "attacking" it, had flipped back on to the seat. Muriel was recalling that the ball was part of the local newspaper. Last week's – and she was also remembering the story she had read in it. How a housewife had been attacked by a wildcat and saved in the nick of time by her dog. The housewife, according to the account, had gone out to lock up her chickens for the night, and had surprised this wildcat who was there obviously to make off with a fowl. The wildcat had turned on the woman, eyes ablaze with ferocity, claws slashing in typical wildcat fashion. It had torn the woman's dress, drawn blood from a shoulder and was about to renew its attack, when the woman's dog, in answer to her screams, dashed to the rescue and drove it off.

Muriel had been anxious that Colin shouldn't read the story. It so happened that on the opposite side of the page there was another news-item which she had felt would interest her family. It provided a sufficient excuse, she felt, to warrant her clipping it and posting it off with her weekly letter; this was how she had been able to explain away the hole in the paper when Colin came to glance at it.

He was becoming increasingly concerned for her; he had warned her several times to watch out against Scrap suddenly turning on her and attacking her. Her claws were sharp enough now – she was coming up for five months old – and she could inflict some really bad scratches, especially if she caught the face. One stab with those flashing claws and she could easily put an eye out. And he had reminded her of Jim Taylor's encounter with a wildcat in the blizzard; and how Gilbert Fisher, who knew about wildcats from personal experience over a long time, even he had been unable to prevent the pair he'd reared from turning into bundles of fury, and in the end had had to put them in Edinburgh Zoo. Muriel couldn't believe it. You just couldn't. This dear, loving, not fully grown kit, forever purring her affection, suddenly launching herself at you like some savage tiger?

And Colin had been forced to admit that he hadn't yet observed the slightest evidence that Scrap was ever anything but as affectionate and friendly as, for example, any domestic cat. To Muriel it seemed impossible that she would ever react in any way other than with total trust and love.

It seemed to her that Scrap was in fact acquiring unique individuality. Everything about her personality, her character, appeared full of a consciousness of Muriel's and Colin's presence. She gave you the sense that, every second she was with you, she fully realised who you were and what you were. Once in a while, you might catch her wearing that odd, abstract expression, as if her thoughts were far away, which made you wish you could reach back with her into that world, the secrets of which were her very own. That world into which you couldn't enter, those secrets you couldn't share. There just weren't the powers of communication between you and her. But it wasn't often; mostly she obviously lived for the here and now to the full joy of it. Never indifferent or aloof when you were with her. The way she walked, the sort of little swaggering way she had. Her attitude, the fun and enjoyment when she played which she shared with you. All seemed to show that she had become a part of the family so to speak. As much as Sanders was, almost. She was no outsider, she had become individualised.

It was as if, Muriel had said to Colin, she realised that he had saved her from death. And that by her attitude to them, and not forgetting Sanders who was also part of the family, she was demonstrating her gratitude. Colin had explained that in fact it was unlikely that Scrap would remember her illness. She wouldn't know that she had been close to death. She knew nothing about death anyway. She would never know that for her as with all living creatures death was inevitable.

But Muriel wondered if, then, that she had forgotten that she had been ill – wouldn't she still remember the warmth and love they had given her? She didn't think this was sentimentalising, or "humanising" her. She thought that Scrap must have built up her character, her nature, her personality – these must have been built on what her senses informed her about the two human beings with whom she had been in close contact these past months, only a short space of time in human estimation but a long long time in Scrap's. The way she had been handled by Colin and Muriel, the way Sanders approached her and sniffed at her with his old nose, the way he allowed her to play games with his tail, all must have impressed themselves upon Scrap so that she had learned to look on life, to scent the life about her, to learn from her tremendous sensory powers exactly what was going on. Surely she was growing closer and closer to the two people to whom she owed not only life itself but her continued welfare, her continued happiness or whatever was the standard of life by which she lived?

Scrap came out from under the seat where she had chased the paper ball, holding it in her mouth. She might have been holding a real prey in her jaws; her expression was enigmatic, she was swinging her head from side to side. Her movements were slow, tensely feline. She looked up, saw Muriel standing there observing her.

She suddenly dropped the ball of paper. After a tiny pause with her eyes flashing she took a long run and then leapt straight up at Muriel.

Chapter Twenty-one

Muriel might have been forgiven if she had reacted with some sign of fear. Scrap's jump at her was so totally unexpected. But as she gained her upper arm, and then reached her shoulder, Muriel didn't even wince. The tiniest gasp of surprise. That was all. Perhaps because it all happened so quickly. But she liked to believe that it was because she had never felt any fear of Scrap.

She remained perfectly still.

Perhaps her heart raced a little, but she didn't turn her face away. Her face, which Scrap could easily have slashed to shreds; but her claws were retracted. For a few moments she held her hands by her sides, then she raised them and held Scrap to her. Back of her mind stirred the remembrance of that elusive intuitive sensation she had experienced, foreshadowing something like this. She was trying to recapture the thought, trying to recall when it had been in her mind, but she couldn't quite bring it back. The purring in her ear as Scrap rubbed her face against hers was quite extraordinary. There was nothing soft, feminine, about it. It was gruff and rough, but full of warmth, charged with affection. It was noisy enough to drown any extra loud beat of Muriel's heart.

She hadn't spoken to her. Only held her, pressing her warm body against her face. She could feel the strong, white whiskers tickling her neck. The warmth of her body was electric, powerfully vibrant; it seemed to fill Muriel's entire world. Scrap clung to her, curling herself round her shoulders tightly so that Muriel made a movement to steady herself.

And then she held her head back and Scrap's face was looking directly into hers. Muriel found herself staring into the golden, amber depths of those eyes with that vertical slit. It was almost a mesmeric stare to which Scrap submitted her, it was as if Muriel was staring right deep into Scrap's eyes, even beyond, deep into her brain. Beyond that even, deep into Scrap's world.

And as for Scrap, what was she seeing?

She wasn't seeing Muriel as, for instance Colin, saw her. You couldn't be certain that Scrap's eyes were just the window of her brain. You don't know that her eyes aren't really an extension of her brain, with which she could select as well as see, even eliminate from the picture, or distort, however her brain directed. It's part of the enigma of Scrap; there's so much you don't know about her. So much that remains a mystery.

But Muriel felt that she was learning something about the world of a wild creature. These were only fleeting moments which she guessed she might never experience again, but during these moments she was thinking that she knew, for the first time, she really knew, what made Scrap tick.

What was really going on behind those brilliant eyes? Inscrutable eyes, because you couldn't really interpret the message they were conveying. You could only believe that she was looking at you, thinking of you with love. Love by her standards. Muriel and Scrap had played games together, she had fondled Scrap with all the affection she could feel for a strange, wild animal about whom she knew so little. She had spoken to her in affectionate tones, she had tweaked her ears and pulled gently at that black striped short tail with its "shaving brush" end. She had demonstrated her love for her, in human terms at any rate. In return Scrap had demonstrated affection, trust and lovingness. In her terms. Muriel had often frowned with puzzlement at what was really going on, when they had played their games, and Scrap had purred and licked her fingers and rubbed her body against her. What had those feline movements really been about, and was Scrap in fact trying to impart some message to her? Was she signalling her and what were the signals?

Both she and Colin knew that they could never enter into Scrap's world any more than she could enter into theirs. Neither would ever solve the other's enigma. It was only that now for those few fleeting seconds, while Scrap held her with that warm golden gaze and she smelled the warmth of her fur, that Muriel felt that even for a brief moment she had received some glimpse of Scrap's secret mysterious world.

It suddenly occurred to her that perhaps it was simply that Scrap was thinking of her as a tree. After all, she was a creature of the forest, instinctively trees represented security to her. And it could be that this was the reason for Scrap's jump at her. Suddenly, spontaneously, Scrap's feeling of security must have been so deep that some force had impelled her to this demonstration of trust.

Muriel smiled at the thought. She felt deeply moved. It was as if a new bond had been forged between her and Scrap.

And then there was Sanders, suddenly awake and barking away, and the spell was broken. Muriel heard Colin's voice, shutting up Sanders, and he stood in the doorway. He stared at her and Scrap. Now, tension filled the caravan. Alarm gave place to curiosity, and then he pushed Sanders aside as the great, black dog, mouth wide and tongue slobbering, tried to welcome him home. Sanders gave a grunt of protest, slanted a narrow look at Scrap, and slumped in his usual corner.

Quietly, Colin shut the door behind him.

Had he arrived in time to save Muriel from an attack by Scrap? He was trying to read the scene. Always at the back of his mind was the fact that you had to be prepared for that soft, feline shape suddenly switching into a bundle of unbridled ferocity. He had warned Muriel, pictured for her Scrap, claws spread, slashing, tearing at her face and eyes. Tearing at her arms and shoulders.

He said: "What about supper, then?"

She looked at him over Scrap's striped hindquarters, saw the reaction in his face and smiled at him. The tension decreased. Colin went slowly over to Muriel, paused for a moment, listening to that noisy purring, and scratched Scrap's neck. Her back arched, her tail was straight upright.

"I think she thinks I'm a tree," Muriel said. She sat down on the long seat, and Scrap duly jumped from her shoulder, ran along the seat, poised above Sanders and dived at him. Sanders grunted protestingly, rolled his eyes at Colin and Muriel, and then Scrap proceeded to lick his face at which he snorted and sprawled over on his side. Scrap turned away, jumped back onto the seat and stared into space.

Colin sat down beside her, holding out his hands. She climbed into his arms, and while Muriel began busily preparing supper, he took her outside to her own home.

After he had locked the gate, he stood for a few moments chatting to her, while she kept up her purring and rubbed her body against the wire-netting. He said goodnight to her and returned to the caravan.

When he had finished supper and was drinking coffee, he said to Muriel: "Of course, nobody would believe it." He would hardly have believed it himself if he hadn't seen it with his own eyes. Scrap curling herself round his wife's shoulders, purring and pushing her face into her hair. Muriel was explaining again, wonder still in her voice, how it had happened.

Colin shoved Sanders' old nose away from his cup of coffee – Sanders had this unfortunate habit of nudging your elbow, to attract your attention, just as you raised your cup to your, mouth, with the result that coffee could splash all over you – and said that he thought he agreed with Muriel. What had happened meant that the bond of affection between her and Scrap had been strengthened. Scrap's sense of security was stronger. Perhaps, though he didn't think much of it, the symbolism of the tree was part of the image she saw.

He still didn't believe, he said. It might not mean that Scrap mightn't yet suddenly revert to her natural wild state. It wouldn't be fair to her to think this was now the case.

She was still a wildcat. Not far off being a full-grown wildcat. Her place was in the wild.

Muriel was looking out of the window in the direction of Scrap's enclosure. It was a dark September evening. Colin opened the door – Sanders immediately concluding he was going to be taken for a swim in the burn, started barking his head off, until told he was out of luck – and Muriel stood by her husband. He put his arm round her shoulder and they stared out.

Scrap will be all right, Colin told Muriel. He smiled at her, a little sadly; he was remembering that night, when Scrap had almost died; and this was the way it came up and could hit you, this emotional involvement with an animal. You needed to know fair and square where your responsibility lay. It lay with the animal. It was Scrap you had to think of. Not your love for her.

Muriel couldn't help wishing that Scrap could stay with them. That she didn't have to go away.

She could still feel her warmth round her shoulders, although she knew it was really Colin's arm. She could still hear that wonderful rough, vibrant purr, although it was Colin, saying that Scrap would be happy.

And that was all that counted.

Chapter Twenty-Two

It was in the fading light of that November afternoon that Colin stopped his car outside Cat Lodge. He and Muriel got out and stood before the tall, iron gates, beyond which the gravel drive ran up to the house. The gates were hung on two great pillars, and on top of each a wildcat stood carved in stone.

The light was failing earlier than Colin had expected – when they had left Kincraig, with Scrap safe in her enclosure and Sanders dozing in the caravan, the afternoon had been bright and sunny, the almost bare branches of the birches shining silver – but now the sky was blackening, where snow thinly edged the profile of Monadhliath Mountain's summit. Colin had brought Muriel to see for the first time Cat Lodge, once famous as an outpost of the Macphersons, when they ruled the moors and glens and the hills of Badenoch from Cluny Castle, hidden by trees about a mile across the moor – Colin and Muriel had glimpsed it on their way to the house.

It had been a difficult place to find, and it was as if you had travelled for miles and miles over this bleak, open moorland, the heatherclad hills rolling away to rise up into the mountains. A grey shrike, its hawk's bill wide open for insects, flew across their path, and then Muriel glanced up to watch a hovering shadow-like buzzard, its wings slanted against the wind as it swung in a wide circle, hunting for prey.

Now you were in the heart of Macpherson country. At the

very heart of the great Clan Chattan terrain; this was wildcat country – the country of the untameable, ferocious wildcat which had symbolised the fearsome Macpherson warriors, who had adopted it for their crest.

And looking at this grey-stoned, low-built house which had known so much of the Highlands' history, bloodshed, war and violence, you could feel it all around you. The atmosphere was charged with mystery and electrical excitement. You could feel it in your bones, it seemed to run through your hair. Here, in the gathering shadows, you might have moved back through the centuries, you might be in the realm of the Clan Chattan, one of the clansfolk yourself.

Across the road from the gates, Colin pointed out a low-built farmhouse. Once it had been the local jail – Cat Lodge itself had served as a courthouse – for a small township had at one time stood here. Now, nothing was left to show it, the houses and crofts had been razed to the ground by marauding bands, or their occupants driven away by the horrors of the '45 or the dreaded Highland Clearances.

Colin pointed to where, past the cross-roads, a narrow road dipped – once a wood had spread beside the road. Gallows Wood, it had been named. Old folk claimed to remember the gibbet that had stood in the middle of the wood, from which malefactors lodged in the jail across the way had been hanged. Now, wood and gibbet and all were gone.

The car's headlights bored through the increasing darkness and Colin and Muriel headed back towards Kincraig. The wind was lifting the light snow on the mountain summit, so that it looked like a ghostly mist. A squall of icy rain splashed against the windscreen. The desolute moor on either side slid past, and every now and again you could catch a glimpse of a lonely house, the glimmer of its lights showing against the darkness.

Muriel's mind was a mix-up of images of that strange house called Cat Lodge, with those two wildcats at the front gates. The savageness of the scene, the violence embodied in the poise of the wildcats, contrasted so vividly with the image of Scrap.

Colin slowed the car. And even as he did so, Muriel saw the flash in the murk ahead, the flash of two greenish brilliantly golden eyes. The eyes glared at you, reflecting the blazing car-lights. It was the first time that Muriel had seen a wildcat in the wild. It kept to the side of the road as the car approached and now Colin was driving alongside it.

Now the wildcat turned and, its thick, bushy tail with its black shaving-brush end held horizontally behind it, ran along the side of the road ahead of the car. Muriel could make out its flat, large head as its body gathered speed. The body was larger than Scrap's and appeared much more muscular. Its hind legs were longer and more powerful. It didn't hurry-scurry like an ordinary cat. It bounded along like a racing greyhound. In the headlights of the car its tawny brindle body striped with black made a dramatic picture, but above all you noted its tail, that short bushy tail with black rings and blunt black end.

"A wildcat all right," Colin muttered. "That tail makes it certain."

He knew that there were feral cats in the neighbourhood, quite a number of them. And he knew, too, that they were often mistaken for wildcats. But no feral cat had ever been known to possess a short black-ended tail like the one that was spread out behind the wildcat racing along ahead of them.

Although it was travelling so fast it had yet never stumbled. All the time its whiskers must have been in sensory contact with every bump, incline and decline in its path. During all of that race along the side of the road its whiskers, pointing downwards, always remained in tactile communication with the ground. The wildcat was always ready for any hollow, stone or bush, or anything that might be in its path, so that it could instantly change course within less than a second.

And this, of course, was the purpose of its tail, it acted as a rudder and received vibrations from the atmosphere about it, from the earth over which it was running, so that it was guiding it all the time in its headlong rush to escape from the oncoming car. You realised now why the tail was held out horizontally behind as it raced along, its tactile sensory "rudder" always in contact with the earth.

By now the car and the wildcat were side by side. Colin looked at his speedometer. It showed that they were travelling at nearly twenty-five miles an hour. That was faster than a human being could run.

Muriel was scared that the wildcat would turn into the road and be run over. But Colin shook his head though he took care to see that the wildcat did not get too far ahead. This is the trouble with instinct, Colin was thinking. The wildcat had learned instinctively over thousands of years to fear man because for thousands of years man had hunted it and killed it. That was why it had developed the instinct to avoid man, because it knew that man meant death, annihilation to it. But the car was relatively recent and the wildcat had not yet learned how to cope with it. That was why instead of just waiting and allowing Colin to pass, leaving it safe to pursue its own way, it panicked and rushed ahead in an effort to outrace the demon behind it.

Then suddenly it turned and disappeared into the blackness of some scrub and bushes by the side of the road. And there the wildcat found its refuge. Muriel gave a sigh of relief that at least the wildcat was safe and Colin drove on.

That night, after supper, Muriel stood at the window and pulling back the curtain she looked in the direction of the children's corner in the park and Scrap's enclosure.

The wind off the loch slapped against the caravan. She could hear the branches of the birches by the glen scrape against each other. She was still haunted by Cat Lodge; she could see it still, mysterious and strange as the daylight faded. And the iron gates, with the grimacing wildcats carved in stone on each pillar.

And, then, once again, she could see the wildcat in the light of the car's headlamps. A living, racing creature, its eyes blazing back at you. Fierce and wild. And free. And entirely different from Scrap.

And yet she knew that they were identical.

It might easily have been Scrap out there on the moorland road, her eyes blazing at you. Scrap, wild and free.

She turned back to Colin who was fondling Sanders with a rough tenderness. And she knew now with a strange kind of conviction that it was right for Scrap to be going to the park, to what was virtually the wild.

Somehow seeing the wildcat in the car's headlights had made her realise where Scrap's real world was.

PART FOUR

Chapter Twenty-three

And so the snow came, and Scrap tasted her first snow-flakes, and turned to watch with concentrated fascination the foot-prints she left behind her as she strutted round the enclosure. She loved the snow; she played games of her own invention in it, pouncing on imaginary prey and ploughing up clouds of it, until she was covered in the stuff from tail to quivering whiskers. Then she would shake herself vigorously, to emerge from her own miniature snow-storm, all spick and span, her tongue licking the last few icy drops off her nose, her brilliant eyes glowing with the fun and excitement of it all.

She would watch Sanders, who always seemed extra frisky on a cold, snowy day; he would burrow his way into a snow-drift until he was right out of sight. Scrap's eyes would glitter as she stared at the heap of snow beneath which Sanders heaved and shoved, until he suddenly appeared, like some huge overgrown mole, puffing and blowing, and spraying snow all around him. In Scrap's enclosure there weren't any snowdrifts into which she could burrow, emulating Sanders, otherwise she would no doubt have followed his example. Nor were there any frozen puddles for her to crack, as Sanders did, and then lie in the icy water crunching up the bits of cracked ice, with a noise that fairly set your teeth on edge to listen to it.

Now the winter solstice passed, the short, grey days lengthened, the whooper swans, those noisiest of swans, which had trumpeted their annual arrival on Loch Insh, were cutting great arrow-heads in the sky above the steely water – soon they would be planning their formation for the flight back north.

It was the second Monday in March that Muriel carried Scrap to the wildcat enclosure in the park, Colin accompanying her, and Sanders limping ahead. Of course, he had known the past couple of days something was in the wind. Every now and then he would start a fit of barking for no reason, and Muriel would have to quieten him with a biscuit. You could say that he was simply barking as a trick to obtain another biscuit, but you should know him better than that. He barked because he knew that a big change was about to be made. A change affecting Colin and Muriel – and himself.

Had Scrap sensed what was to happen to her?

She hadn't appeared in any way anxious about her future; when she was with Muriel in the caravan she still purred happily and appeared as loving as ever. Was she any different in the way she acted to Colin? You couldn't say she was. Her attitude towards him was as friendly and affectionate as it had always been.

But she must have picked up some information for herself from Sanders' barking and uneasiness. You couldn't imagine that somehow she hadn't been alerted. Only she wasn't giving anything away.

She was aged ten months; she had grown into a big strong-looking wildcat, with her tiger-like stripes, each one almost totally black now against their background of thick, golden tawny fur. She might grow larger, she wouldn't finally stop growing for at least another year. But her illness might have affected her in this respect, might have cut her back from ever becoming as full-sized as a wildcat in the wild.

All the same she was big enough, sinuously muscled, compactly powerful, and very beautiful.

The dusky haze of evening was just beginning to fall as Colin and Sanders and Muriel with Scrap reached the wildcat enclosure. There was a barrier between the public and the strong wire-netting which surrounded the enclosure.

Colin removed the heavy padlock from the gate and Muriel and Scrap followed him inside. He told Sanders to wait, and the dog gave a typical growl of disappointment – he would have very much enjoyed an exploratory sniff round – and loped off a few yards, where he stretched himself out to watch what was going to happen through half-closed eyes.

Muriel put Scrap down. She and Colin watched the wildcat. For a few moments she stood there, and then began to walk around in circles sniffing the air. A light breeze came over from the surrounding hills which were still glimmering white at the snow-line. The light was beginning to fade. Scrap gave a sudden jump and ran off for several yards. She ran lightly and happily, her tail erect, and then suddenly she stopped as if frightened of the freedom with which she suddenly found herself.

Colin and Muriel remained, watching, while behind them Sanders had scrambled to his feet and limped over to note more closely what was going on. There was a sudden electricity in the air, as if Scrap was going to do something to celebrate her fredom, as if she might perhaps undergo some wild paroxysm. But instead she turned and faced Colin and Muriel and walked towards them.

She stopped and sniffed the air again. She came forward very slowly and Colin and Muriel could hear her loud throaty purring.

"She's going to be all right," Colin said.

It was his idea that Scrap should be the first to occupy the enclosure before the other expected wildcats so that she should be able to settle down. He was looking forward to news that a gamekeeper had caught a wildcat for him. Perhaps one with her kits – this was the time when wildcats were most vulnerable and liable to be caught. Soon Scrap would have companionship, perhaps even a mate. One day she might even have kits of her own.

Colin turned on his heel and went towards the gate. Muriel

was watching Scrap as she herself stood there watching them.

"Goodnight, Scrap," she said. She followed her husband. As she looked back Scrap was still standing there and they could still hear her purring on the breeze.

Colin followed Muriel out, then he closed the gate behind him and locked the padlock carefully. They stood watching Scrap for several more moments. She made no move towards them; just stood there and they could see her eyes glinting as the dusk deepened.

Muriel was very silent as she and Colin turned away, and they went slowly towards the caravan with Sanders following them for a few moments before he rushed ahead, barking.

What was Scrap thinking, left alone there in her new territory?

The sky was darkening above her and a star here and there was beginning to appear. She could sniff the scent of the snow on the summits of the Monadhliath Mountains which lay to the north and she could also sniff the sea from Moray Firth beyond, which was where Loch Ness drained into the sea. As well as these scents all about her she could smell the ducks and geese from the tiny loch down in the nearby glen. She could scent the foxes in their enclosure just across the way, and the scent which she had never known before, of the wolves where they lived on the rocky ridge nearby.

There were many new scents, strange scents that she had never sniffed before. The grass under her feet, the undergrowth, the scent of the trees. And there were those other scents that she was missing, of Muriel and Colin and Sanders.

Did she feel lonesome where she was in that wide space to which she had never before been accustomed? Or did the sudden freedom thrill her? Did the wildness of her blood race through her so that she revelled in her new freedom? Instinctively she played games, she went through the motions of hunting, though she would perhaps never kill anything because there would always be food for her provided by Colin. And there was a pool specially built.

Scrap reached the top of the hillock above the rocky cave which Colin had carefully prepared for her as a form of shelter. Her lair. Inside was a rough sandy floor with plenty of strong rock and earth; it was roomy and comfortable.

On the top of the hillock Scrap faced north and tasted a snowflake or two which had drifted over from the mountains. Her paws disturbed some loose scree and she listened to it as it scattered down over the rocks to come to rest against an edge of rough grasses. She tapped some more scree deliberately with her paw and listened to it as once again it scattered down the side of the hillock to come to rest as before. As she considered the movement of the bits of scree and the debris and listened to it rattle down her attention was interrupted by the hoot of a tawny owl. She heard an answering hoot from the owl's mate and she watched the pale ghostly shape as it flew noiselessly overhead. She could see that its eyes were saucers of curiosity. She knew that the owl had seen her down below but it did not hesitate in its flight. Scrap was too big, too powerful-looking to offer any hopes of becoming a prey to the tawny owl's talons.

Scrap turned away and moved slowly, carefully down towards her lair. She was purring to herself, a deep throaty purr. Her bushy tail was held upright as she made her way jauntily towards the cave.

Next morning, she was in her lair when she caught a scent she knew and heard a voice. "Scrap."

She went out quickly to see Muriel approaching. She ran down to her, purring. Muriel couldn't open the gate into the enclosure, because Colin had the only key. She knelt by the gate, pushing her fingers through the wire-netting, and Scrap purred as she was scratched under her chin, and she licked Muriel's fingers, and pushed her body against the wire-netting, trying to get as close to Muriel as she could.

Chapter Twenty-four

Muriel awoke suddenly. It was just past two a.m. She glanced at the window, moonlight shafted through a gap in the drawn curtains. It would be a bright sky. Even as she awoke she was thinking of Scrap, and guessed she had probably been dreaming about her.

She had visited her regularly, as had Colin, the last three weeks since she moved to the park. Scrap had quickly acclimatised herself to her new surroundings, eating the food that was left her, and she was looking really beautiful, her fur all glossy, her wonderful golden eyes bright and full of happiness.

On an impulse, without realising what she was doing, Muriel slipped out of bed as quietly as possible, hoping not to disturb her husband. She pulled on her shoes and slacks, put on her raincoat, took a torch and, managing not to wake Sanders, reached the caravan-door. As she turned the handle, Sanders turned over with a heavy flop, growled, and for a moment she feared she had awoken him. He may even have opened one eye at her, but she didn't wait to see; she hurried out, closing the door gently behind her.

She hardly needed the torch, the moonlight was so bright. She reached the enclosure. She could see no sign of Scrap. She called to her and there was no reply.

For a chilling moment fear assailed her as she wondered if perhaps something had happened to Scrap. She switched on the torch and the light flickered over the enclosure. And then suddenly a pair of eyes were blazing at her from out of the shadows.

"Scrap," she called softly.

A miaow and then that rough gruff purring note as Scrap ran towards her. She pushed herself against the wire-netting, arching her back, shaving-brush tail stuck upright, while Muriel scratched the top of her head with a finger. Scrap responded with that throbbing purr-purr. Then suddenly she drew away. The purring stopped. She was staring past Muriel, her eyes ablaze, her ears swivelled forward.

"Scrap – "

Muriel started. It was Colin's voice, very quiet, behind her. She swung round; her heart stopped racing as she saw him there.

He had heard Muriel go out of the caravan, and guessed that she had wanted to know how Scrap was getting on in the loneliness of the night. Scrap had scented him and, even as she did so, she'd heard the ticking of his wrist-watch.

Colin opened the enclosure-gate, and Muriel went in and sat down, her raincoat protecting her from the wet grass, while Scrap came to her. She fondled her, talking to her quietly. Colin stood there looking down at them, then he moved away shining his torch in the direction of Scrap's lair. It looked comfortable enough, and Scrap had quickly made it her home.

Several moments Muriel sat there, holding Scrap in her arms, her soft fur against her cheek and the feel of those electric whiskers; Scrap's body throbbed with her deep purring as if a motor engine were inside it.

She tweaked Scrap's ears. Her remarkable ears which she could swivel about so effectively. It might be a clever, individual trick for you to wiggle your ears, but it was part of the structure of Scrap's that she could wiggle hers, move them around until they were pointing in the direction of the sound she was registering. Her ears were even larger than they appeared.

Muriel could feel growing round their base the fur, which concealed their true size – and which helped keep them warm and protect the internal ear from harmful dirt or bits of debris.

Colin moved closer, and Scrap twitched her ears. She seemed to tense. She backed against Muriel – and then jumped from her arms, took two or three paces towards Colin, then suddenly crouched and took that fantastically quick run and jumped.

She gained Colin's shoulder and was curving her body round his neck, purring and rubbing her face against his. Muriel had caught her breath, and started to call out to warn Colin.

But like her, on that earlier occasion, he didn't wince. He felt no fear that this was a sudden attack. As Scrap ran up his shoulder, he held her firmly, gently, and muttered words of endearment. This was the first time that Scrap had jumped at him in this way.

He knew – and Muriel watching them together, also knew – that despite the fact that she had gone back to the wild to which she belonged Scrap would always remain his and Muriel's friend. She would always love them both and just so long as they came to see her she would always come to see them.

A few minutes later, with Muriel beside him, Colin locked the padlock on the enclosure-gate. They looked back and they could see Scrap staring after them. They called out goodnight to her and then turned to go home, and as they did so they heard Sanders barking. In a few moments his lumbering shape came limping towards them, still barking, then at a word from Colin he quietened down and, pausing to give Scrap a wuff over his shoulder, walked back beside them.

Afterword

Scrap's extraordinary and fascinating story teaches us – every one of us, and certainly all Macphersons – another side of the wildcat, more of its ways, its gentleness, as well as its courage and its fighting spirit to survive, more than could ever have been guessed before.

The widely varying accounts of the wildcat's life in the wild are well known, and many myths and legends are attached to the animal. To read Scrap's unique story is to make us all wonder if enough is really known about her kind. Perhaps the time has come when the wildcat should be regarded less as a pest to be trapped, poisoned, shot and exterminated at all cost, as vermin, and more as worthy of preservation and respect as part of Badenoch's history and its natural life.

Surely we must agree for certain that these beautiful creatures must be symbolic not of Man's taking from nature, his destruction of our wild animals, but of returning to nature, Man giving back to the wild his love?

Long may the wildcat live on in the beautiful wilds of the Highlands that are its natural home.

William Macpherson of Cluny
27th Chief of the Clan Macpherson